Nano and Microsensors for Chemical and Biological Terrorism Surveillance

Nano and Microsensors for Chemical and Biological Terrorism Surveillance

Edited by

Jeffrey B.-H. Tok

Lawrence Livermore National Laboratory, Livermore, CA, USA and Micropoint Biosciences Inc., Sunnyvale, CA, USA

RSC Publishing

ISBN: 978-0-85404-140-4

A catalogue record for this book is available from the British Library

©Royal Society of Chemistry, 2008

Published by The Royal Society of Chemistry,
Thomas Graham House, Science Park, Milton Road,
Cambridge CB4 0WF, UK

Registered Charity Number 207890

For further information see our website at www.rsc.org

Preface

The 9/11 attack on US soil has inadvertently heightened the need for our preparedness in other potential means of terrorist attack. In particular, both biological and chemical warfare have been at the top of the priority list of most governmental agencies as reagents can be covertly prepared and disseminated to result in both widespread fear and casualties. Among many others, one primary preventive step in preparing for the above attacks is to establish a network for efficient surveillance and rapid detection such that appropriate response to such attacks can be timely and effective.

Over the years, primarily due to technological advances, both chemical and biological agents that are able to inflict mass destructions have become more diverse and complex. Subsequently, improvement of sensing devices for rapid and sensitive detection should also be made to keep pace with these engineered or emerging threat agents. Advances in micro- and nanofabrication techniques to enable sensing devices are especially of interest as they have been shown to offer desired advantages such as improved and enhanced functionality, increased efficiency and speed in their readout, reduction in their fabrication cost, and also reduced reagent consumption. Indeed, numerous innovative and exciting reports which took advantage of the above-mentioned techniques for both chemical and biological sensing have appeared over the last decade. While it is not the intention of this book to detail each reported approach, the aim is to compile in depth several detection schematics such that the reader can be provided with a general sense of these micro- and nanoscale sensing systems and platforms.

In this book, I have assembled a series of chapters detailing both well-established and "next-generation" micro- and nanoscale sensors and/or sensing platforms. Briefly, these sensors or sensing platforms range from the novel utilization of nanotubes, cantilevers, nano- and/or microsized pores and engineered whole cells to polymeric transistors *etc.* for sensing purposes. It is truly gratifying to see a synergistic marriage of myriad techniques, ranging from

Nano and Microsensors for Chemical and Biological Terrorism Surveillance
Edited by Jeffrey B.-H. Tok
© Royal Society of Chemistry, 2008
Published by the Royal Society of Chemistry, www.rsc.org

chemical, engineering and biological, for the development of sensors, which
was once traditionally thought to be reserved for immunologists. The enabling
of the above technologies should soon result in a much improved sensing
network for the detection and surveillance of both chemical and biological
warfare agents.

Lastly, I thank the various members in my research group, namely Hansang
Cho, Nick Fischer, Eric Schopf and Aaron Rowe, for their help in the com-
pletion of this book project.

Jeffrey B.-H. Tok
Livermore, California

Contents

Nano and Microsensors for Chemical and Biological Terrorism Surveillance
Edited by Jeffrey B.-H. Tok
© Royal Society of Chemistry, 2008
Published by the Royal Society of Chemistry, www.rsc.org

**Chapter 3   Resistive-pulse Sensing and On-chip Artificial Pores
                for Biological Sensing**
                Omar A. Saleh and Lydia L. Sohn

**Chapter 4   Micro- and Nanocantilever Systems for
                Molecular Analysis**
                Sibani Lisa Biswal

Chapter 7   Whole-cell Sensing Systems in Chemical and Biological Surveillance
Elisa Michelini, Luca Cevenini,
Laura Mezzanotte and Aldo Roda

Chapter 8   Conducting Polymer Transistors for Sensor Applications
Fabio Cicoira, Daniel A. Bernards and George G. Malliaras

Contributors

Matthew J. Aernecke, *Tufts University, Department of Chemistry, 62 Talbot Ave, Medford, MA 02155, USA*

Sarit S. Agasti, *Department of Chemistry, University of Massachusetts, 710 North Pleasant Street, Amherst, MA 01003, USA*

Daniel A. Bernards, *Department of Materials Science and Engineering, Bard Hall, Cornell University, Ithaca 14850, USA*

Sibani Lisa Biswal, *Department of Chemical and Biomolecular Engineering, Rice University, MS 362, 6100 Main Street, Houston, TX 77005, USA*

Luca Cevenini, *Department of Pharmaceutical Sciences, University of Bologna, Via Belmeloro 6, 40126 Bologna, Italy*

Fabio Cicoira, *Department of Materials Science and Engineering, Bard Hall, Cornell University, Ithaca 14850, USA and IFN-CNR, via alla Cascata 56/c, 38050 Povo (Trento), Italy*

Keith E. Herold, *Department of Bioengineering, University of Maryland, College Park, MD 20742, USA*

George G. Malliaras, *Department of Materials Science and Engineering, Bard Hall, Cornell University, Ithaca 14850, USA*

Laura Mezzanotte, *Department of Pharmaceutical Sciences, University of Bologna, Via Belmeloro 6, 40126 Bologna, Italy*

Elisa Michelini, *Department of Pharmaceutical Sciences, University of Bologna, Via Belmeloro 6, 40126 Bologna, Italy*

Avraham Rasooly, *FDA Center for Devices and Radiological Health and NIH-National Cancer Institute, 6130 Executive Blvd. EPN, Rockville, MD 20852, USA*

Aldo Roda, *Department of Pharmaceutical Sciences, University of Bologna, Via Belmeloro 6, 40126 Bologna, Italy*

Vincent M. Rotello, *Department of Chemistry, University of Massachusetts, 710 North Pleasant Street, Amherst, MA 01003, USA*

Omar A. Saleh, *Materials Department and Biomolecular Science and Engineering Program, University of California, Santa Barbara, Santa Barbara, CA 93106-5050, USA*

Eric S. Snow, *Institute for Nanoscience, Naval Research Laboratory, Washington, DC 20375, USA*

Lydia L. Sohn, *Department of Mechanical Engineering, University of California, Berkeley, Berkeley, CA 94720-1740, USA*

David R. Walt, *Tufts University, Department of Chemistry, 62 Talbot Ave, Medford, MA 02155, USA*

Chang-Cheng You, *Department of Chemistry, University of Massachusetts, 710 North Pleasant Street, Amherst, MA 01003, USA*

CHAPTER 1

Carbon-Nanotube-Network Sensors

ERIC S. SNOW

Institute for Nanoscience, Naval Research Laboratory, Washington, DC 20375, USA

1.1 Introduction

The growing threat of chemical, biological and radiological attack has created a demand for sensors that are capable of monitoring a large number of facilities for the preemptive detection or potential release of toxic agents. Such applications are highly demanding, requiring inexpensive sensors that are extremely sensitive while producing a low incidence of false alarms. Many such applications are beyond the capability of current technology, which has motivated the development of improved chemical and biological sensors.

Nanomaterials, because of their intrinsically high surface-to-volume ratio, offer the potential to advance the state of the art by serving as the active material for chemical, biological, radiological and explosive sensors. Among such nanomaterials single-walled carbon nanotubes (SWNTs) possess a number of intrinsic properties that make them particularly well suited for a wide range of sensor applications. SWNTs are single-atomic sheets of graphite rolled into a cylinder ~ 1 nm in diameter that can range in length from 10s of nanometers to 100s of microns depending on the method of growth and preparation.[1-3] Because SWNTs are composed entirely of surface atoms, molecular adsorbates can significantly perturb their electronic properties.[4,5] SWNTs also exhibit near-ballistic electron transport along the tube axis,[6] which provides a highquality electrical conduit for the transmission of such electrical perturbations to external contacts. Finally, the graphitic surface of SWNTs is chemically robust, enabling long-term stable operation.

Nano and Microsensors for Chemical and Biological Terrorism Surveillance
Edited by Jeffrey B.-H. Tok
© Royal Society of Chemistry, 2008
Published by the Royal Society of Chemistry, www.rsc.org

Initial laboratory results demonstrated the capability for SWNTs to electronically detect the adsorption of chemical and biological analytes.[4,7,8] However, a number of significant scientific and technological challenges inhibited the transition of these demonstrations to commercial sensor technology. These challenges include the development of an inexpensive, high-yield nanotube device fabrication process, addressing the high level of low-frequency noise, and achieving analyte specificity. Researchers have made significant strides at addressing each of these problems enabling the commercialization of SWNT sensor technology.

In this chapter we examine the current state of development of carbon nanotube chemical and biological sensors. Such sensors can take several forms, which include electrochemical sensors,[9–12] ionization sensors[13] and field-effect transistors (FETs)[14,15] with the SWNT FET platform perhaps the most developed of these. Each of these sensor platforms has its particular set of device physics, design issues and application areas, and it would be difficult to thoroughly discuss each of these in a limited space. Consequently, this chapter will focus on the SWNT FET used for the direct electronic detection of gases, chemical vapors and biological analytes. This chapter is divided into four sections, which include sensor design and fabrication, electronic transduction and noise, chemical vapor and gas detection, and biological detection. These topics cover the main areas of SWNT-FET-sensor research and development. For the interested reader, excellent reviews exist in the literature of other nanotube-based sensor platforms.[10–12]

1.2 Sensor Design and Fabrication

Initial demonstrations of the sensor properties of SWNTs were performed on FETs that contained a single SWNT as the conducting channel (see Figure 1.1).[4,5] In such devices the SWNT was grown or deposited on the surface of a thermal oxide on a conducting Si substrate. Metal source/drain electrodes formed the electrical contacts, and the Si substrate served as a back gate. Such devices were instrumental in investigating the charge-transfer properties of molecular adsorbates and in demonstrating the potential of SWNTs for sensor applications. However, such single-nanotube devices are not easily manufactured, because it is difficult to precisely position individual SWNTs, since the variation in SWNT electronic type (due to diameter and chirality variations[3]) produces large device-to-device non-uniformity, and because individual SWNTs produce a high level of low-frequency noise.[16–19] Consequently, factors such as these have impeded the commercialization of single-SWNT FET sensors.

1.2.1 SWNT Networks

A practical solution to the fabrication problem consists of fabricating field-effect transistors in which the conducting channel is composed of a SWNT random network.[20] SWNT networks are two-dimensional arrays of randomly

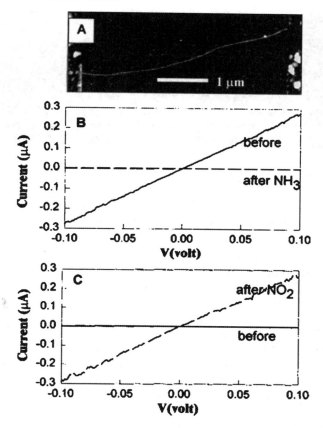

Figure 1.1 (A) Atomic-force-microscope image of a SWNT FET. Current-voltage characteristics recorded before and after exposure to NH_3 (B) and NO_2 (C). For (C) the current *versus* voltage curves were recorded under a gate bias of $+4\,V$. Reproduced with permission from [4]. Copyright 2000 American Association for the Advancement of Science.

positioned SWNTs (Figure 1.2A). If the density of SWNTs in the channel is sufficient that they highly intersect then the SWNTs form an electrically continuous film over arbitrarily large dimensions. Sensors formed from such networks are inexpensive to manufacture using conventional microfabrication techniques and exhibit uniform properties that reflect the aggregate properties of many random, individual SWNTs.[21] The networks are typically grown directly on the thermal oxide of a Si substrate or deposited onto a substrate from solution. Under the appropriate conditions SWNT networks with sheet resistances typically between 10 and 1000 kΩ/square can be grown or deposited uniformly across the surface of large-area substrates.[22]

A key to the electronic properties of SWNT networks is the electrical contact that is formed between intersecting nanotubes lying on a surface. SWNTs adhere to surfaces *via* van der Waals forces.[23] Because SWNTs are extremely

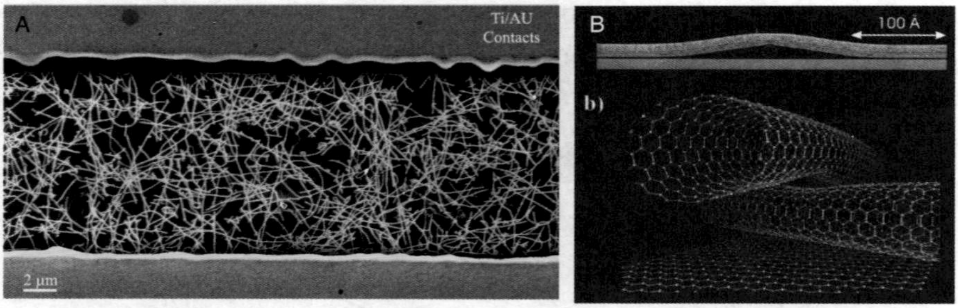

Figure 1.2 (A) Atomic-force-microscope image of a SWNT network FET. (B) Simulation of two intersecting SWNTs lying on a surface. The van der Waals forces acting on the top SWNT are sufficient to deform the SWNTs at the point of intersection. (B) reproduced with permission from [23]. Copyright 1998 the American Physical Society.

stiff (Young's modulus $\sim 10^{12}$ Pa),[24] when two SWNTs cross the van der Waals force pulling down on the top SWNT is transferred to the point of intersection. This force is sufficient to deform the two SWNTs forcing them closer together than the interplane spacing in graphite (see Figure 1.2B).[23] This close contact increases the inter-nanotube tunneling probability, which in the case of two metallic SWNTs can be as high as $0.1\,e^2/h$[25] (where $4\,e^2/h$ is the ideal ballistic conductance of a SWNT). Metal-semiconductor inter-SWNT contacts result in a higher resistance caused by the Schottky barrier formed between the two SWNTs.[25] Such electrical point contacts between intersecting SWNTs create an electrically continuous network over arbitrarily large dimensions, provided that the level of interconnectivity exceeds the percolation threshold for conductivity. Such films can range from semiconducting to metallic behavior depending on the density of SWNTs and the device geometry.[20,26]

It should be noted that recently the Rogers group at the University of Illinois has demonstrated that highly ordered arrays of SWNTs can be grown on certain substrates (see Figure 1.3).[27,28] If the cost of such ordered arrays can be kept sufficiently low it may be possible to manufacture sensors with precisely aligned SWNTs that avoid any deleterious effects of the inter-nanotube contacts present in a network. This approach offers promise for significant improvement in SWNT-sensor performance.

1.2.2 Sensor Fabrication

Sensors consist of microfabricated metal electrodes deposited on a patterned SWNT network that is typically formed on the thermal oxide of a conducting Si substrate.[29] The device structure is that of a thin-film transistor with a back gate that is formed by the Si substrate. A schematic of a sensor is shown in Figure 1.4. For biosensing the sensor is sometimes submerged in a saline solution that contains a Pt electrode used as an electrochemical gate.[30] SWNT

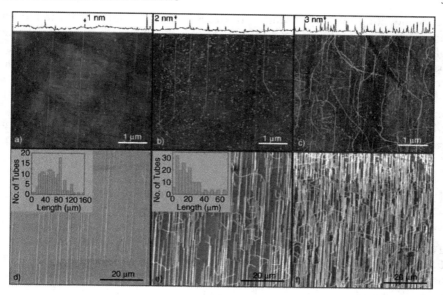

Figure 1.3 (a)–(c) AFM images of aligned SWNTs grown on single-crystal quartz substrate using different densities of catalyst particles. (d)–(f) Large-area SEM images of tubes grown in this fashion. These results indicate a decreasing degree of alignment with increasing tube density. Reprinted with permission from [28]. Copyright 2005 John Wiley and Sons, Inc.

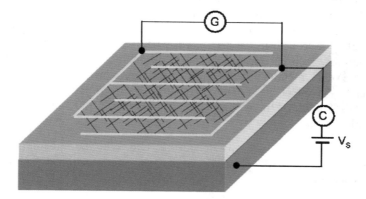

Figure 1.4 Schematic of a SWNT network FET sensor. A conducting Si substrate, separated from the network by a layer of SiO_2, serves as a back gate. Molecular adsorption on the SWNTs is detected as a change in the network conductance and/or the network capacitance. Reprinted with permission from [29]. Copyright 2005 American Chemical Society.

network sensors are simple to fabricate, and the design exposes the surface of the SWNTs to the environment for efficient molecular detection.

For both electronic and sensor applications, an important issue is the role of nanotube/metal contacts. Metal electrodes form a Schottky barrier to semiconducting SWNTs,[31] which introduces a series resistance that can reduce sensitivity. This contact resistance can be minimized by the appropriate choice of metal, *e.g.* Pd[6], and/or annealing the contacts,[32] and using large spacing between the electrodes such that the network resistance is much larger than the contact resistance.

For biosensing researchers have demonstrated that binding events at or near such Schottky contacts can significantly modulate the contact resistance,[33] which produces an additional sensor response. In this case, the electrode design can be optimized to either enhance or suppress such contact effects.

1.3   Electronic Transduction and Noise

1.3.1   Conductance

The conductance, G, is measured between the source and drain electrodes of the FET, and the substrate electrode forms a capacitive link, C_{NET}, to the SWNT network. An applied substrate voltage, V_s, produces a charge, $C_{NET}V_s$, on the network that can be used to modify the interaction with certain analyte molecules[34] or to calibrate the charge transfer from adsorbates.[29] An AC voltage applied to the substrate can be used to directly measure C_{NET}, which is typically $\sim 10\,nF/cm^2$. Thus, C_{NET} for a $1\,mm^2$ sensor is $\sim 100\,pF$, which is easily measured to high accuracy.

The adsorption of analyte molecules can have a number of effects on the electrical properties of a SWNT FET sensor. Charge transfer between molecular adsorbates and the SWNTs can affect the channel conductance by changing the number of mobile charge carriers.[4] Adsorbates can also produce scattering centers that lower the carrier mobility.[35] Additionally, molecular adsorption can affect the SWNT/metal-contact Schottky barrier, which modifies the contact resistance[36,37] or affect the tunneling transmission from the metal contacts into the SWNTs.[38] Each of these effects causes a change in the conductance of the SWNT FET.

1.3.2   Capacitance

In addition to the conductance effects, the fringing fields emanating from the SWNTs can polarize adsorbed molecules.[39] Thus, the dielectric properties of adsorbates can affect C_{NET}, which provides a second, independent method of detecting the presence of molecular adsorbates.[29] Measurements on devices with a 100-nm-thick oxide indicate a comparable conductance and capacitance response for many adsorbates. However, capacitance detection has the advantage of significantly reduced low-frequency noise (see Section 1.3.3 below), which leads to lower detection limits for comparable levels of sensitivity.[40]

In addition to the dielectric effects, charge transfer from adsorbates can also produce a capacitance response. The SWNT network capacitance can be modeled as two capacitors in series, $C_{NET} = (1/C_G + 1/C_Q)^{-1}$ where C_G is the geometric gate capacitance and C_Q is the quantum capacitance, where C_Q reflects modifications to the SWNT network Fermi energy, E_F, in response to changes in charge.[41] At zero temperature, $C_Q = e^2 g(E_f)$, where $g(E)$ is the SWNT network density of states (for finite temperature thermal broadening effects have to be taken into effect to calculate C_Q[42]). Adsorbate charge transfer can shift E_f into a region with a different density of states resulting in a change in C_Q, which produces a capacitance response.

For vapor or gas detection C_G is typically much smaller than C_Q (10 *vs.* 100 aF/µm of SWNT), and $C_{NET} \approx C_G$. In this case the capacitance response is dominated by dielectric effects for most adsorbates.[29] However, for biodetection the SWNT FET is sometimes operated with an ionic solution gate. The electrolyte double-layer produces a large C_G, which can be larger than C_Q,[41] and $C_{NET} \approx C_Q$. In this case charge effects can dominate the capacitance response.

1.3.3   Electrical Noise

For sensor applications an attractive feature of nanoscale materials such as SWNTs is that they possess a high surface-to-volume ratio, which can produce a high sensitivity. However, the gains in sensitivity can be offset by high noise levels unless appropriate measures are taken in the sensor design. Of particular importance is low-frequency noise, since chemical and biological detection is typically performed at frequencies <10 Hz. At these low frequencies the dominant type of noise is $1/f$, which is universally present in all electronic systems. Because the amplitude of $1/f$ noise varies inversely with number of charge carriers in a system,[43] nanoscale materials (which by definition have a small number of charge carriers) produce a large value of low-frequency noise. Thus, the construction of nanoscale sensors degrades the signal-to-noise and the corresponding detection limits.

In order to retain the high sensitivity of nanoscale materials and produce a low-noise device, the sensor can be constructed using many nanoscale components to increase the total number of charge carriers. For this reason, SWNT-network FETs have much lower detection limits than individual-SWNT FETs due to the much improved signal-to-noise ratio.

It has been empirically established that the $1/f$-noise-power density, $S_V(f) = \alpha_H/N\ V^2/f$,[43] where V is the applied voltage, N is the number of charge carriers in the system and α_H is the Hooge parameter, which for bulk electronic materials is typically of the order of 10^{-3}.[44] This equation indicates that nanoscale devices, because of their small value of N, will exhibit a large component of low-frequency noise. Indeed, large $1/f$ noise has been noted in many studies of nanotube devices.[17,19,45–50] Surprisingly, studies indicate that the Hooge parameter for SWNTs is comparable to that of bulk electronic materials, even though SWNTs consist entirely of surface atoms.[16,18] Thus, single-SWNT devices are noisy, not because of poor material quality or surface effects, but because the devices contain an extremely small number of charge carriers.

One solution to the noise problem is to construct relatively large-area sensors using many SWNTs, *i.e.* N $\propto$ area.[48] According to the Hooge formula, the magnitude of the sensor $1/f$ noise, equal to $(\int S_V(F)dF)^{1/2}$, will decrease in proportion to the square root of the area of the sensor; while its sensitivity will be area independent (assuming that the number of detected analytes is proportional to the sensor area). Thus, the signal-to-noise ratio will increase as the square root of the area of the sensor.

SWNT networks provide a convenient means to incorporate millions of SWNTs in a single device in order to improve the signal-to-noise ratio. By using a gate bias to calibrate the charge sensitivity of SWNT networks, it has been established that network sensors can detect in conductance a charge perturbation $\sim 0.01\,e^-$ per μm of SWNT,[15] which is the same order of charge transfer obtained from a single molecular adsorption event.[51] Thus, properly designed SWNT networks can achieve single-molecule-per-SWNT detection sensitivity.

The use of capacitance-based detection can further improve the signal-to-noise ratio. Unlike resistors, ideal capacitors do not produce $1/f$ noise. Ideally, SWNT network capacitors would be noise free (limited only by the noise of the measurement circuit); however, there is a small component of $1/f$ noise in SWNT capacitors due to the contribution of the quantum capacitance. Because C_Q is sensitive to charge, the same charge fluctuations that produce conductance noise will introduce fluctuations in the measured capacitance. However, as described above in the absence of an electrolytic gate the capacitance is much less sensitive to charge than the conductance, so the level of $1/f$ noise is proportionally reduced. The capacitance noise-power density is typically about a factor of ~ 1000 times smaller than the conductance noise (see Figure 1.5).[40] Given that the signal levels for capacitance and conductance sensing are comparable, the key advantage of capacitance sensing is that the signal-to-noise ratio is further improved. Thus, the decreased level of $1/f$ noise is a key advantage of capacitance-based sensing for SWNT FETs.

1.4 Gas/Vapor Detection

Kong *et al.*[4] and Collins *et al.*[5] were the first to establish that certain molecular adsorbates can significantly alter the electrical conductance of SWNTs. In such cases charge transfer between the absorbate and the SWNT causes either an increase or a reduction in the number of mobile charge carriers. Kong *et al.*[4] demonstrated that NH_3, an electron donor, causes a reduction in the conductivity of p-type semiconducting SWNTs while NO_2, an electron acceptor, causes an increase in conductance (see Figure 1.1). Subsequent studies have shown that a large number of gases and chemical vapors measurably alter the electrical properties of SWNTs, both in conductance and capacitance, with sub-part-per-billion sensitivity obtainable for certain analytes. This capability for highly sensitive detection of molecular adsorbates makes SWNT FETs an attractive platform for the detection of both permanent gases and chemical vapors.

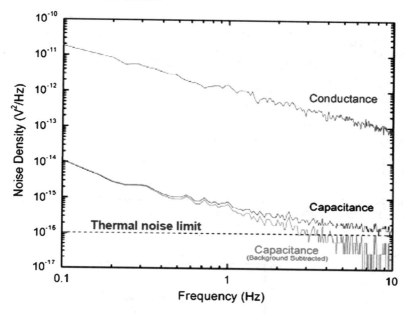

Figure 1.5 Comparison of the low-frequency conductance and capacitance noise of a SWNT-network FET sensor. Note that the low-frequency capacitance noise density is about three orders of magnitude smaller than the conductance noise measured in the same device. Reprinted with permission from [40]. Copyright 2007 Elsevier Ltd.

1.4.1 Gas Detection

While SWNTs have demonstrated the potential to detect a variety of permanent gases such as HCN, Cl_2, HCl, CO, CO_2, *etc.*[35,52–57], perhaps the most studied gases are NH_3 and NO_2. NH_3 and NO_2 both produce large, partially recoverable changes in the electrical properties of SWNTs. In both cases the adsorbate binding energy is sufficiently large that the adsorbate remains attached to the SWNT long after the analyte is removed from the surrounding atmosphere. Typically, heat or ultraviolet light is required to desorb the analyte and return the SWNT to its initial conductance value,[55,58,59] although other means have been developed.[60] This long desorption time causes the SWNT to act like a dosimeter integrating the dose of analyte. In this way, SWNTs can detect long-term exposure to extremely low doses of these gases, *e.g.* SWNT sensors coated with polyethyleneimine (PEI), which enhances the sensitivity to NO_2, respond in about 1000 s to concentrations as low as 100 parts per trillion (see Figure 1.6).[7]

Theoretical studies of the interaction of SWNTs with molecular adsorbates indicate a strong interaction between SWNTs and NO_2 with significant charge transfer, but only a weak interaction with NH_3.[51,61] This latter result is at odds with the strong conductance response and thermal desorption data, which indicate that the most of the bound NH_3 desorbs at high temperature.[62]

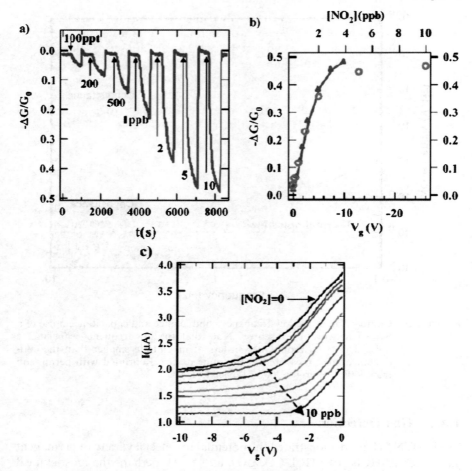

Figure 1.6   (a) Change in conductance normalized by initial conductance (G_0) at $V_g = 0$ as a function of time for a PEI-functionalized n-type MT device exposed to various concentrations of NO_2 gas. The device was exposed to each concentration of NO_2 for 10 min, after which recovery was made by UV light (254 nm) desorption of NO_2. The concentration of NO_2 was varied by diluting 100 ppm of NO_2 (in Ar) with air by using four mass-flow controllers. The diluted gas was then flown into a homemade chamber that houses the sensor chip. (b) Conductance change *vs.* concentration of NO_2 (top horizontal axis, data points in circles) and conductance *vs.* gate-voltage (bottom axis, data points in triangles) respectively for an n-type MT device. (c) I-V_g curves for a device recorded after exposure to NO_2 of successively increasing concentrations (from black curve to blue curve: $[NO_2] = 0$, 0.1, 0.2, 0.5, 1, 2, 5, 10 ppb respectively). Reprinted with permission from [7]. Copyright 2003 American Chemical Society.

This discrepancy between experiment and theory has caused researchers to consider other more stable binding sites such as structural defects. Andzelm *et al.* studied interaction of a number of SWNT defects with NH_3.[63] They found that defects like vacancies and oxygenated Stone-Wales defects spontaneously chemisorb NH_3 into dissociated NH_2 and H. The strong binding energies ($\sim 2.5\,eV$) and charge transfer ($\sim 0.1\,e$) are in better agreement with experiment. This result indicates that structural defects may play a significant role in the sensor properties of SWNTs for other analytes as well.

1.4.2 Vapor Detection

In addition to gases SWNT FETs have demonstrated the capability to detect a large number of organic and inorganic vapors including simulants for chemical agents and explosives.[7,34,39,40,53,55,64–67] Unlike NH_3 and NO_2, for many of the vapors the response and recovery is fast ($<1\,s$), allowing real-time detection (see Figure 1.7). However, because the molecules desorb rapidly, exposure to trace-level concentrations produces only a small fraction of a monolayer

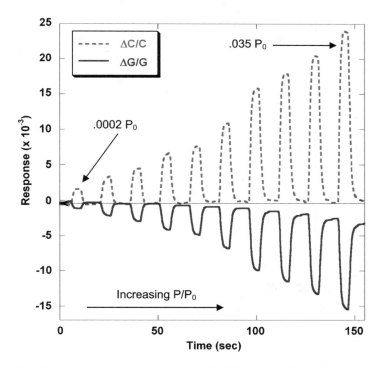

Figure 1.7 The relative capacitance response ($\Delta C/C$) and relative conductance response ($\Delta G/G$) for a SWNT FET exposed to 5-s pulses of acetone at varying dilutions of the saturated vapor. Reprinted with permission from [40]. Copyright 2007 Elsevier Ltd.

coverage of analyte on the SWNT surface. This low surface coverage, combined with the fact that physisorbed molecules produce $\ll 1$ electron of charge transfer, results in small changes in the SWNT conductance. Consequently, minimizing noise in order to maximize the signal-to-noise ratio is of paramount importance for real-time trace-level detection.

For this reason, C-based detection with its low level of low-frequency noise can provide the necessary ratio of signal-to-noise. Robinson *et al.* have shown that C-based detection can detect a large number of chemical vapors, with projected detection limits of simulants for a chemical agent (dimethyl methylphosphonate) and explosive (dinitrotoluene) below 1 ppb.[40] Thus, real-time, reversible detection of sub-ppb of toxic and explosive vapors is achievable with SWNT FET sensors using the capacitance mode of transduction (see Figure 1.8).

The rapid recovery from exposures of many chemical vapors implies a weak physisorption process, which is indicative of the interaction predicted for molecules physisorbed on the pristine SWNT surface.[51] However, recent data indicate that structural defects are important even in the case of these weakly interacting vapors. Robinson *et al.* have shown that the introduction of a small

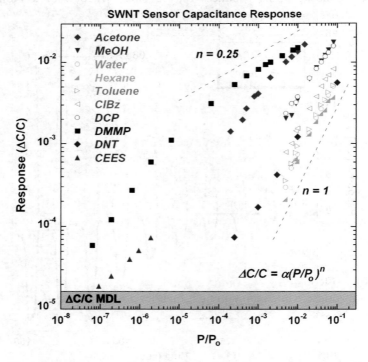

Figure 1.8   Relative capacitance response ($\Delta C/C$) measured as a function of concentration in response to doses of various chemical vapors. The concentration is reported in units of the saturated vapor pressure, P_0. The dashed horizontal line at the bottom of the figure represents the minimum detectable level (MDL) based on a signal-to-noise of 3.

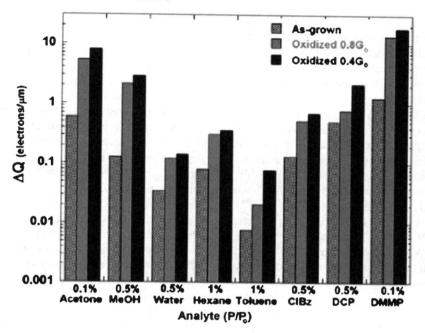

Figure 1.9 Charge transfer in units of electrons per μm of SWNT measured for various vapors before and after oxidation of the SWNTs. The introduction of structural defects *via* oxidation increases the amount of measured charge transfer for all of the analytes. Reprinted with permission from [68]. Copyright 2006 American Chemical Society.

number of structural defects ($\sim 1\%$) can increase the sensitivity of a SWNT FET sensor by more than an order of magnitude (see Figure 1.9).[68] This large increase is due to the increased binding energy at such defects compared to binding to the pristine SWNT surface.

This observation that a small number of binding sites can control the sensor response has an implication for using covalent functionalization to introduce chemical selectivity. If covalent functionalization of the majority of the SWNT surface were required for selectivity, then it would be difficult to introduce selectivity without severely degrading the electrical properties of the SWNTs. The results of Robinson *et al.* indicate that large changes to the sensor response can be introduced while maintaining a high level of SWNT conductivity.

1.4.3 Chemical Specificity

The demonstrated ability for SWNT sensors to detect trace levels of a wide range of chemical vapors and gases facilitates a range of applications in the detection of chemical agents, explosives and other toxic chemicals. However, a necessary requirement for a detection system is that it should be capable of detecting trace levels of the analytes of interest in the presence of background concentrations of potential interfering vapors.

It is extremely difficult to produce a sensor that responds exclusively to a single chemical vapor or gas, so approaches to isolating the response from potential interferents have been designed. One approach is to construct an "electronic nose" that is composed of an array of sensors with each array element functionalized to preferentially respond to a particular class of chemical.[69] An analysis of the array response is then used to identify an unknown vapor and to exclude the response from interferents. In order to impart a degree of selectivity to array elements several approaches have been applied, which include coating with polymers,[15,39,70,71] grafting polymers to SWNTs,[66,72–74] covalent functionalization,[65,68] coating with metal nanoclusters[7,56,75–77] and attaching strands of DNA.[67,78,79]

Lu *et al.* have constructed such a nose using a thirty-two-element array of SWNT FETs consisting of pristine SWNTs, polymer-coated SWNTs, and SWNTs doped with metals[52] (see Figure 1.10). The array was exposed to NO_2, HCN, HCl, CL_2, acetone and benzene, and a pattern recognition technique was applied to the results. The sensor array was able to discriminate each of the vapors and gases based on their chemical nature at part-per-million concentration levels. While such a nose approach is promising, analysis of the array response is quite complicated when the detection system is presented with three or more vapors simultaneously.

An alternative and potentially more powerful approach is to construct a detection system with a micro-fabricated front-end delivery system that preferentially concentrates the analyte(s) of interest and then separates them from background interferents.[80] In this approach the ambient air is first passed through a chemoselective preconcentrator that selectively sorbs the analyte of interest and then releases the collected vapor in a short pulse by rapidly heating the preconcentrator. The resulting pulse of analyte, which still may contain trace interferents, is then passed through a microfabricated gas chromatograph to separate the remaining vapor components in time in order to isolate the analyte of interest. The separated stream of air is then presented to a sensor (or an array of chemoselective sensors), which indicates whether any analyte is present in the appropriate time window. In principle, such a system can deal with complex ambients, although at the cost of increased response time due to the time delay caused by the preconcentration and gas chromatography. Such a micro gas chromatograph has been constructed using SWNTs as the stationary phase in order to achieve rapid separation for faster analysis times.[81]

1.5 Detection of Biological Agents Using Immunoassays and DNA Hybridization

A promising application of SWNT FETs is the label-free electronic detection of bio-recognition events such as antibody/antigen interactions and DNA

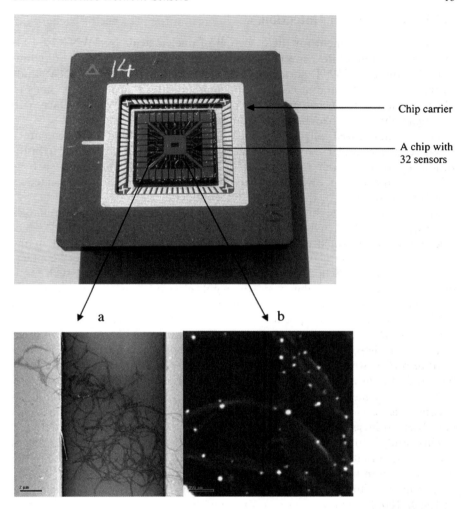

Figure 1.10 A schematic of a silicon-based sensor array chip with two representative images of nanomaterials in the interdigitated electrode platform: (a) a SEM image of pristine carbon nanotubes across two electrodes, (b) a SEM image of carbon nanotubes loaded with monolayer protected clusters of gold nanoparticles. Reprinted with permission from [52]. Copyright 2006 Elsevier Ltd.

hybridization. The SWNT-FET sensor provides an inexpensive platform for the rapid, direct electronic detection of such bio-recognition events, which requires no external modification such as labeling of the analyte biomolecules. In this section we discuss the use of SWNT FETs functionalized with single-stranded DNA (ss-DNA), antibodies and aptamers for direct electronic biological detection.

1.5.1 DNA-based Detection

A necessary requirement for the detection of nucleic acids by SWNT FETs is the ability to immobilize specific sequences of nucleic acids such as oligonucleotides on the SWNT surface. Experiments indicate that ss-DNA binds strongly to SWNTs, and molecular modeling indicates that the DNA binds to SWNTs due to the nucleic acid-base π-π stacking on the SWNT surface with the sugar-phosphate backbone extending away from the nanotube.[82]

These immobilized oligonucleotides can selectively bind complementary DNA sequences under a hybridization state, although the kinetics of hybridization are much slower than free solution DNA.[83] Since processes such as DNA hybridization involve charge interactions, such events can be detected as a direct perturbation of the SWNT FET conductance.

Although it is expected that SWNT FETs should respond directly to the charges associated with DNA hybridization, there is some uncertainty as to the exact mechanism of electronic detection. The hybridization can either result in the charging of the SWNT network or in the modulation of the SWNT-metal contacts *via* a change in the Schottky-barrier height. Tang *et al.* found that such modulation of the energy level alignment with a gold electrode can be a primary source of the conductance modulation.[84] They found that this transduction mechanism is in many ways superior to optical and electrochemical methods of detection.

Star *et al.* demonstrated that hybridization on ss-DNA-functionalized SWNTs can detect the presence of single nucleotide polymorphism.[85] 17-mer capture probes were used to detect a single-base-pair mismatch between wild-type and mutant alleles (see Figure 1.11). SWNT network FET transfer characteristics were measured dry, before and after hybridization. A significant drop in conductance was observed in devices exposed to matched ss-DNA, which was not observed in sensors exposed to ss-DNA with a single-base mismatch.

This study reported two additional significant findings. The researchers demonstrated that the single nucleotide polymorphism could be detected in the presence of non-homologous DNA. In order to accomplish this non-specific adsorption was inhibited by treating the SWNT network with a blocking agent, Triton X-100, following functionalization with the capture probe. Importantly the physisorption of the blocking agent did not displace the capture probe. Additionally the researchers demonstrated that DNA hybridization could be greatly enhanced by the addition of $MgCl_2$ salt to the buffer solution. The presence of Mg^{2+} ions greatly enhanced the efficiency of the hybridization, leading to a 1000-fold increase in detection sensitivity. The addition of Mg^{2+} increased the sensitivity from 1 nM to 1 pM for a one-hour incubation time.

Gui *et al.* have demonstrated that further enhancements in sensitivity can be achieved by the addition of a threading intercalator.[86] The researchers used an intercalator, which binds strongly to double-stranded DNA. The intercalator contained a redox-active ligand $Os(bpy)_2Cl^+$ that caused a reduction in the SWNT network conductivity (see Figure 1.12).

As an alternative detection approach the SWNT network can be used with a charge sensor that is capacitively coupled to a DNA-functionalized gate

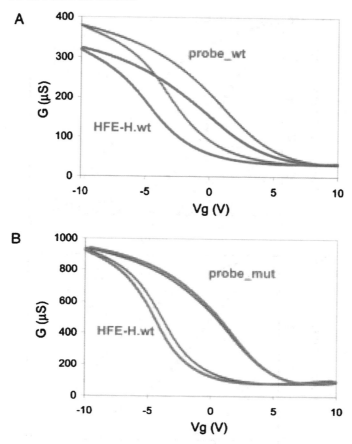

Figure 1.11 Electronic detection of the presence of a single nucleotide polymorphism. (A) $G-Vg$ curves after incubation with allele-specific wild-type capture probe and after challenging the device with wild-type target (50 nM). (B) $G-Vg$ curves in the experiment with mutant capture probe containing a single-base mismatch. Reprinted with permission from [85]. Copyright 2006 National Academy of Sciences.

electrode.[87] In this example the SWNTs are not directly exposed to the DNA. Instead peptide nucleic acid oligonucleotides were covalently immobilized to a gold back-gate electrode. Hybridization with complementary DNA or RNA sequences caused a negative surface charge on the electrode, which produced an increased conductance in the p-type SWNT network. The detection of concentrations of DNA down to 6.8 fM was reported. However, experiments with non-complementary DNA showed similar results, which was attributed to non-specific binding to the Au back-gate electrode. Comparisons between the complementary and non-complementary results indicated an increased response for the complementary sample.

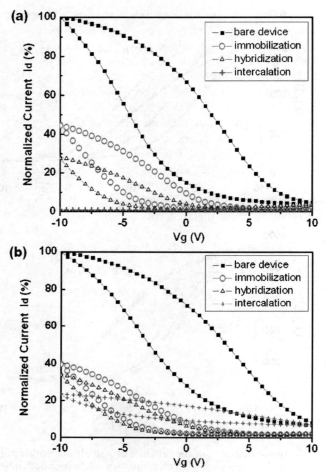

Figure 1.12 Typical gate voltage dependence of the normalized drain current *Id* normalized by the initial drain current of their bare device at $V_g = -10\,\text{V}$ for (a) a CNNFET bare device immobilized with NH_2-DNA, hybridized with complementary target analyte, (b) a CNNFET bare device, immobilized with NH_2-DNA, hybridized with single-base mismatched target analyte, and immersed with the PIND-Os intercalator source-drain bias was kept at $-0.5\,\text{V}$. Reprinted with permission from [86]. Copyright 2006 American Institute of Physics.

1.5.2 Antibody/Antigen-based Detection

SWNT FETs can be configured as a sensitive transducer to electronically detect antibody-antigen interactions, without the need for labeling. In this scheme, the SWNT network is functionalized with an antibody that is used to bind a specific antigen, and an additional blocking agent is used to inhibit the non-specific binding of other proteins to unfunctionalized regions of the SWNT surface.

A variety of proteins non-specifically bind to the SWNT surface.[30,88–93] The resulting adsorption leads to changes in the electronic properties of the SWNT measured either in solution or after subsequent drying.[93] Like DNA-hybridization sensing, there is some uncertainty in the mechanism causing the conductivity change. The conductivity changes have been attributed to charge transfer arising from interactions of $-NH_2$ groups of the protein with SWNT surface.[93] Alternatively, Chen *et al.* selectively blocked protein attachment either on the SWNTs or on the contacts and found that electronic effects at the metal/SWNT contacts contribute significantly to the electronic signal, *e.g.* due to changes in the metal work function, which modulates the SWNT-electrode Schottky barrier.[33]

In order to achieve specific detection and to avoid electrical detection of non-specific adsorption on the SWNT surface requires both the functionalization with the specific antibody of interest and coating of the unfunctionalized SWNT surface with a blocking agent that prevents all other non-specific binding. Star *et al.* demonstrated such specific binding and the associated electrical detection of streptavidin binding to biotin.[94] In this example, the authors measured transfer characteristics of a SWNT FET dry, before functionalization, after functionalization and after exposure to matched and unmatched proteins. The authors coated the SWNTs with a combination of polyethyleneimine (PEI) and poly (ethylene glycol) (PEG). The PEI was used to attach biotin for the specific recognition of streptavidin. The PEG was used as a blocking agent for non-specific binding to the portions of the SWNT uncoated with PEI (see Figure 1.13).

With this combination the authors observed a large drop in conductivity upon exposure to streptavidin consistent with an introduction of scattering

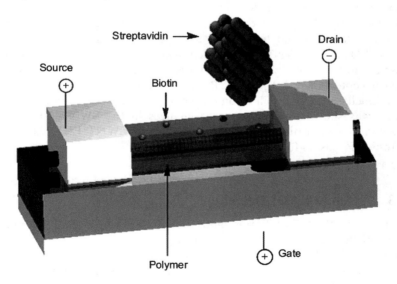

Figure 1.13 Schematic of a SWNT FET biosensor. A polymeric functional layer, which coats the nanotube, is functionalized with a molecular receptor, biotin, a protein that recognizes a biomolecule, streptavidin. Reprinted with permission from [94]. Copyright 2003 American Chemical Society.

centers, not an electrostatic gating effect. The authors also blocked the biotin binding sites and demonstrated that, in this case, the adsorption and subsequent detection of streptavidin was blocked, indicating that non-specific adsorption was effectively blocked by the polymers.

Chen et al. used a slightly different approach to functionalization and were able to detect the antibody/antigen recognition events in a buffered saline solution.[30] In this case, the SWNT network was first coated with a blocking agent, e.g. Tween 20. The blocking agent was activated with 1,1-carbonyldi-imidazole, which is then conjugated with specific receptors. The Tween 20 was shown to be an effective blocking agent for the non-specific adsorption of protein, while the specific binding to the receptors produced a drop in conductivity. Tests with SWNTs functionalized with antigens demonstrated that the antigens retain their ability to bind to their respective antibody with a high degree of specificity. The SWNT network FETs were used to detect antibodies associated with human autoimmune diseases at concentrations below the threshold for fluorescence-based detection (see Figure 1.14).

More recently, Briman et al. have demonstrated a capacitance-based detection technique that was able to detect a specific protein in calf serum.[95] They monitored the SWNT network/solution capacitance as controlled amounts of prostate-specific antigen (PSA) were added to calf serum. In this case the SWNT network was functionalized by incubating the device with an antihuman PSA monoclonal antibody solution. The authors were able to detect PSA in calf serum at concentrations down to 1 ng/mL. In this case the capacitance change can arise from changes either in the ionic solution double layer or in the quantum capacitance of the SWNTs.

Most recently Zhang et al. demonstrated the use of a ligand-receptor-protein system covalently bound to SWNTs as a proof of concept for SWNT biosensors.[96] Oxidized SWNTs were functionalized using amidation with either the Knob protein domain from adenovirus serotype 12 and or with its human cellular receptor, the CAR protein. Both were shown to retain their bioactivity and specificity after immobilization on the SWNT surface. In addition, protein binding was shown to produce large changes in the SWNT FET transfer characteristics, indicating that such functionalized SWNT FETs can serve as electronic biosensors for detecting adenoviruses.

1.5.3 Aptamer-based Detection

In order to use SWNT FETs to monitor in real time biorecognition events, the binding of the analyte must produce a charge in an ionic solution that can be detected by the SWNTs. This is made difficult due to electrostatic screening effects by ions in the solution within a Debye length of the bound charge. This screening length is $\sim 0.32 I^{-1/2}$ nm where I is the ionic strength of the water, which is ~ 1 nm for typical 100 mM ionic solutions.[97] This screening length is much smaller than typical antibodies (~ 10–15 nm), which reduces the charge response of antibody-antigen biorecognition events.

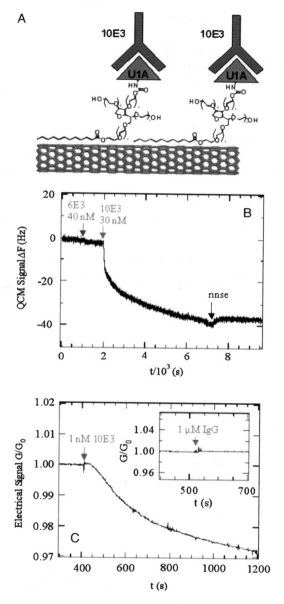

Figure 1.14   Specific detection of mAbs binding to a recombinant human autoantigen. (A) Scheme for specific recognition of 10E3 mAb with a nanotube device coated with a U1A antigen–Tween conjugate. (B) Quartz crystal monitor frequency shift *vs.* time curve showing selective detection of 10E3 while showing rejection of the antibody 6E3, which recognizes the highly structurally related autoantigen TIAR. (C) Conductance *vs.* time curve of a device shows specific response to ≤ 1 nM 10E3 while rejecting poly-clonal IgG at a much greater concentration of $1\,\mu M$ (inset). Reprinted with permission from [30]. Copyright 2003 National Academy of Sciences.

A promising solution to this problem is the use of SWNTs functionalized with aptamers. Aptamers are synthetic oligonucleotides that are generated to recognize proteins, drugs and amino acids with high specificity using combi- natorial libraries for selective absorption, recovery and amplification.[98,99] For charge-based detection aptamers provide a distinct advantage over antibodies because their small size allows binding of the analyte within a Debye length of the SWNT surface (see Figure 1.15).

Maehashi *et al.* used aptamer-functionalized SWNTs to detect in solution immunoglobulin E (IgE) at concentrations down to 250 pM.[97] They observed a drop in the network conductance over a period of a few minutes following the introduction of IgE (see Figure 1.16). A similar device functionalized with an

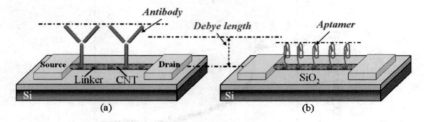

Figure 1.15   Schematic representation of label-free protein biosensors based on CNT-FETs: (a) antibody-modified CNT-FET; (b) aptamer-modified CNT-FET. Reprinted with permission from [97]. Copyright 2007 American Chemical Society.

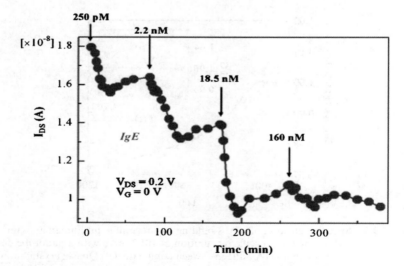

Figure 1.16   Time dependence of source-drain current of the CNT-FET at the source-drain bias of 0.2 V and at the gate bias of 0 V after the introduction of target IgE at various concentrations onto the IgE aptamer-modified CNT-FET. Arrows indicate the points of IgE injections. Reprinted with permission from [97]. Copyright 2007 American Chemical Society.

anti-IgE monoclonal antibody and exposed to IgE resulted in a much smaller, barely observable change in conductance. Thus, the use of aptamer-modified SWNTs facilitates the real-time in-solution detection of proteins.

1.6 Summary

Following the initial demonstrations of sensitivity to molecular adsorbates, researchers directed a significant effort toward the development of SWNT-based sensors. This effort has resulted in the development of the SWNT network FET as a powerful sensor platform with application for both chemical and biological sensing. These sensor capabilities derive from the remarkable properties of SWNTs, which include their unique structure, the near-ideal electron transport, their compatibility with conventional microprocessing, and their practically inert chemical structure, which allows stable operation in many ambient conditions.

The fact that SWNT-based sensors have advanced from the initial laboratory demonstrations in 2000 to an inexpensive, manufacturable sensor platform in the span of just a few years is a good indicator that such sensors will find real-world application in the growing number of commercial, department-of-defense and homeland-security needs. Companies such as Nanomix, Inc. are pioneering the commercialization of the SWNT-based sensors and have established wafer-scale manufacturing capability. Undoubtedly, the on-going research and development efforts on SWNT sensors at university and government research labs will lead to new discoveries, significant improvements in performance and additional areas of application.

Acknowledgements

The author gratefully acknowledges financial support from the Federal Aviation Association, the Office of Naval Research, and the NRL Institute for Nanoscience.

Acronyms

SWNT	Single-walled carbon nanotube
PEI	Polyethyleneimine
MDL	Minimum detectable level
DNA	Deoxyribonucleic acid
ss-DNA	Single-stranded DNA
RNA	Ribonucleic acid
PEG	poly(-ethylene glycol)
PSA	Prostate-specific antigen
MT	Multi-tube

SEM   Scanning electron microscope
CNN   Carbon nanotube network
PIND-Os   N,N'-bis(3-propylimidazole)-1,4,5,8-napthalene diimide modified
             with two Os(2,2'-bipyridine)$_2$Cl$^+$ pendants
IgE   Immunoglobulin E
IgG   Immunoglobulin G

References

 1. S. Iijima, *Nature*, 1991, **354**, 56–58.
 2. S. Iijima and T. Ichihashi, *Nature*, 1993, **363**, 603–605.
 3. H. J. Dai, *Surface Science*, 2002, **500**, 218–241.
 4. J. Kong, N. R. Franklin, C. W. Zhou, M. G. Chapline, S. Peng, K. J. Cho and H. J. Dai, *Science*, 2000, **287**, 622–625.
 5. P. G. Collins, K. Bradley, M. Ishigami and A. Zettl, *Science*, 2000, **287**, 1801–1804.
 6. A. Javey, J. Guo, Q. Wang, M. Lundstrom and H. J. Dai, *Nature*, 2003, **424**, 654–657.
 7. Q. F. Pengfei, O. Vermesh, M. Grecu, A. Javey, O. Wang, H. J. Dai, S. Peng and K. J. Cho, *Nano Letters*, 2003, **3**, 347–351.
 8. J. N. Wohlstadter, J. L. Wilbur, G. B. Sigal, H. A. Biebuyck, M. A. Billadeau, L. W. Dong, A. B. Fischer, S. R. Gudibande, S. H. Jamieson, J. H. Kenten, J. Leginus, J. K. Leland, R. J. Massey and S. J. Wohlstadter, *Adv. Mater.*, 2003, **15**, 1184.
 9. J. S. Ye and F. S. Sheu, *Current Nanoscience*, 2006, **2**, 319–327.
10. N. Sinha, J. Z. Ma and J. T. W. Yeow, *Journal of Nanoscience and Nanotechnology*, 2006, **6**, 573–590.
11. M. Trojanowicz, *TrAC: Trends in Analytical Chemistry*, 2006, **25**, 480–489.
12. J. Li, J. E. Koehne, A. M. Cassell, H. Chen, H. T. Ng, Q. Ye, W. Fan, J. Han and M. Meyyappan, *Electroanalysis*, 2005, **17**, 15–27.
13. A. Modi, N. Koratkar, E. Lass, B. Q. Wei and P. M. Ajayan, *Nature*, 2003, **424**, 171–174.
14. B. L. Allen, P. D. Kichambare and A. Star, *Adv. Mater.*, 2007, **19**, 1439–1451.
15. E. S. Snow, F. K. Perkins and J. A. Robinson, *Chem. Soc. Rev.*, 2006, **35**, 790–798.
16. Y. M. Lin, J. Appenzeller, J. Knoch, Z. H. Chen and P. Avouris, *Nano Letters*, 2006, **6**, 930–936.
17. Y. M. Lin, J. Appenzeller, Z. H. Chen and P. Avouris, *Physica E: Low-Dimensional Systems & Nanostructures*, 2007, **37**, 72–77.
18. M. Ishigami, J. H. Chen, E. D. Williams, D. Tobias, Y. F. Chen and M. S. Fuhrer, *Appl. Phys. Lett.*, 2006, **88**, 203116.
19. P. G. Collins, M. S. Fuhrer and A. Zettl, *Appl. Phys. Lett.*, 2000, **76**, 894–896.
20. E. S. Snow, J. P. Novak, P. M. Campbell and D. Park, *Appl. Phys. Lett.*, 2003, **82**, 2145–2147.

21. E. S. Snow, J. P. Novak, M. D. Lay, E. H. Houser, F. K. Perkins and P. M. Campbell, *Journal of Vacuum Science & Technology B*, 2004, **22**, 1990–1994.

22. J. P. Edgeworth, N. R. Wilson and J. V. Macpherson, *Small*, 2007, **3**, 860–870.

23. T. Hertel, R. E. Walkup and P. Avouris, *Phys. Rev. B*, 1998, **58**, 13870–13873.

24. M. M. J. Treacy, T. W. Ebbesen and J. M. Gibson, *Nature*, 1996, **381**, 678–680.

25. M. S. Fuhrer, J. Nygard, L. Shih, M. Forero, Y. G. Yoon, M. S. C. Mazzoni, H. J. Choi, J. Ihm, S. G. Louie, A. Zettl and P. L. McEuen, *Science*, 2000, **288**, 494–497.

26. S. Kumar, N. Pimparkar, J. Y. Murthy and M. A. Alam, *Appl. Phys. Lett.*, 2006, **88**, 123505.

27. S. J. Kang, C. Kocabas, T. Ozel, M. Shim, N. Pimparkar, M. A. Alam, S. V. Rotkin and J. A. Rogers, *Nature Nanotechnology*, 2007, **2**, 230–236.

28. C. Kocabas, S. H. Hur, A. Gaur, M. A. Meitl, M. Shim and J. A. Rogers, *Small*, 2005, **1**, 1110–1116.

29. E. S. Snow and F. K. Perkins, *Nano Letters*, 2005, **5**, 2414–2417.

30. R. J. Chen, S. Bangsaruntip, K. A. Drouvalakis, N. W. S. Kam, M. Shim, Y. M. Li, W. Kim, P. J. Utz and H. J. Dai, *Proc. Natl. Acad. Sci. USA*, 2003, **100**, 4984–4989.

31. P. Avouris, J. Appenzeller, R. Martel and S. J. Wind, *Proc. IEEE*, 2003, **91**, 1772–1784.

32. Y. F. Chen and M. S. Fuhrer, *Phys. Rev. Lett.*, 2005, **95**, 236803.

33. R. J. Chen, H. C. Choi, S. Bangsaruntip, E. Yenilmez, X. W. Tang, Q. Wang, Y. L. Chang and H. J. Dai, *J. Am. Chem. Soc.*, 2004, **126**, 1563–1568.

34. T. Someya, J. Small, P. Kim, C. Nuckolls and J. T. Yardley, *Nano Letters*, 2003, **3**, 877–881.

35. A. Star, T. R. Han, V. Joshi, J. C. P. Gabriel and G. Gruner, *Adv. Mater.*, 2004, **16**, 2049–2052.

36. V. Derycke, R. Martel, J. Appenzeller and P. Avouris, *Appl. Phys. Lett.*, 2002, **80**, 2773–2775.

37. Y. F. Chen and M. S. Fuhrer, *Nano Letters*, 2006, **6**, 2158–2162.

38. C. Y. Lee, S. Baik, J. Q. Zhang, R. I. Masel and M. S. Strano, *J. Phys. Chem. B*, 2006, **110**, 11055–11061.

39. E. S. Snow, F. K. Perkins, E. J. Houser, S. C. Badescu and T. L. Reinecke, *Science*, 2005, **307**, 1942–1945.

40. J. A. Robinson, E. S. Snow and F. K. Perkins, *Sensor Actuator Phys.*, 2007, **135**, 309–314.

41. S. Rosenblatt, Y. Yaish, J. Park, J. Gore, V. Sazonova and P. L. McEuen, *Nano Letters*, 2002, **2**, 869–872.

42. J. Guo, S. Goasguen, M. Lundstrom and S. Datta, *Appl. Phys. Lett.*, 2002, **81**, 1486–1488.

43. F. N. Hooge, *Phys. Lett. A*, 1969, **A29**, 139–140.

44. F. N. Hooge, *IEEE Trans. Electron Dev.*, 1994, **41**, 1926–1935.
45. F. Liu, K. L. Wang, D. H. Zhang and C. W. Zhou, *Appl. Phys. Lett.*, 2006, **89**, 063116.
46. F. Liu, K. L. Wang, D. H. Zhang and C. W. Zhou, *Appl. Phys. Lett.*, 2006, **89**, 243101.
47. S. Reza, Q. T. Huynh, G. Bosman, J. Sippel-Oakley and A. G. Rinzler, *J. Appl. Phys.*, 2006, **100**, 094318.
48. E. S. Snow, J. P. Novak, M. D. Lay and F. K. Perkins, *Appl. Phys. Lett.*, 2004, **85**, 4172–4174.
49. S. Soliveres, A. Hoffmann, F. Pascal, C. Delseny, M. S. Kabir, O. Nur, A. Salesse, M. Willander and J. Deen, *Fluctuation and Noise Letters*, 2006, **6**, L45–L55.
50. S. Soliveres, J. Gyani, C. Delseny, A. Hoffmann and F. Pascal, *Appl. Phys. Lett.*, 2007, **90**, 082107.
51. J. J. Zhao, A. Buldum, J. Han and J. P. Lu, *Nanotechnology*, 2002, **13**, 195–200.
52. Y. J. Lu, C. Partridge, M. Meyyappan and J. Li, *J. Electroanal. Chem.*, 2006, **593**, 105–110.
53. J. Li, Y. J. Lu, Q. Ye, M. Cinke, J. Han and M. Meyyappan, *Nano Letters*, 2003, **3**, 929–933.
54. O. Kuzmych, B. L. Allen and A. Star, *Nanotechnology*, 2007, **18**, 375502.
55. L. Valentini, C. Cantalini, I. Armentano, J. M. Kenny, L. Lozzi and S. Santucci, *Diam. Relat. Mater.*, 2004, **13**, 1301–1305.
56. Y. J. Lu, J. Li, J. Han, H. T. Ng, C. Binder, C. Partridge and M. Meyyappan, *Chem. Phys. Lett.*, 2004, **391**, 344–348.
57. R. K. Roy, M. P. Chowdhury and A. K. Pal, *Vacuum*, 2005, **77**, 223–229.
58. R. J. Chen, N. R. Franklin, J. Kong, J. Cao, T. W. Tombler, Y. G. Zhang and H. J. Dai, *Appl. Phys. Lett.*, 2001, **79**, 2258–2260.
59. H. Q. Nguyen and J. S. Huh, *Sensor Actuator Chem.*, 2006, **117**, 426–430.
60. T. Zhang, S. Mubeen, E. Bekyarova, B. Y. Yoo, R. C. Haddon, N. V. Myung and M. A. Deshusses, *Nanotechnology*, 2007, **18**, 165504.
61. A. Ricca and C. W. Bauschlicher, *Chem. Phys.*, 2006, **323**, 511–518.
62. M. D. Ellison, M. J. Crotty, D. Koh, R. L. Spray and K. E. Tate, *J. Phys. Chem. B*, 2004, **108**, 7938–7943.
63. J. Andzelm, N. Govind and A. Maiti, *Chem. Phys. Lett.*, 2006, **421**, 58–62.
64. J. P. Novak, E. S. Snow, E. J. Houser, D. Park, J. L. Stepnowski and R. A. McGill, *Appl. Phys. Lett.*, 2003, **83**, 4026–4028.
65. M. L. Y. Sin, G. C. T. Chow, G. M. K. Wong, W. J. Li, P. H. W. Leong and K. W. Wong, *IEEE Transactions on Nanotechnology*, 2007, **6**, 571–577.
66. H. C. Wang, Y. Li and M. J. Yang, *Sensor Actuator Chem.*, 2007, **124**, 360–367.
67. C. Staii and A. T. Johnson, *Nano Letters*, 2005, **5**, 1774–1778.
68. J. A. Robinson, E. S. Snow, S. C. Badescu, T. L. Reinecke and F. K. Perkins, *Nano Letters*, 2006, **6**, 1747–1751.
69. T. A. Dickinson, J. White, J. S. Kauer and D. R. Walt, *Trends Biotechnol.*, 1998, **16**, 250–258.

70. S. Kim, H. R. Lee, Y. J. Yun, S. Ji, K. Yoo, W. S. Yun, J. Y. Koo and D. H. Ha, *Appl. Phys. Lett.*, 2007, **91**, 093126.
71. J. Li, Y. J. Lu and M. Meyyappan, *IEEE Sensor J.*, 2006, **6**, 1047–1051.
72. Y. F. Zhang, C. H. Suc, Z. F. Liu and J. Q. Li, *J. Phys. Chem. B*, 2006, **110**, 22462–22470.
73. E. Bekyarova, M. Davis, T. Burch, M. E. Itkis, B. Zhao, S. Sunshine and R. C. Haddon, *J. Phys. Chem. B*, 2004, **108**, 19717–19720.
74. L. Niu, Y. L. Luo and Z. Q. Li, *Sensor. Actuator. Chem.*, 2007, **126**, 361–367.
75. B. K. Kim, N. Park, P. S. Na, H. M. So, J. J. Kim, H. Kim, K. J. Kong, H. Chang, B. H. Ryu, Y. M. Choi and J. O. Lee, *Nanotechnology*, 2006, **17**, 496–500.
76. R. Larciprete, S. Lizzit, L. Petaccia and A. Goldoni, *Appl. Phys. Lett.*, 2006, **88**, 243111.
77. M. Penza, G. Cassano, R. Rossi, M. Alvisi, A. Rizzo, M. A. Signore, T. Dikonimos, E. Serra and R. Giorgi, *Appl. Phys. Lett.*, 2007, **90**, 173123.
78. A. T. C. Johnson, C. Staii, M. Chen, S. Khamis, R. Johnson, M. L. Klein and A. Gelperin, *Semicond. Sci. Tech.*, 2006, **21**, S17–S21.
79. A. T. C. Johnson, C. Staii, M. Chen, S. Khamis, R. Johnson, M. L. Klein and A. Gelperin, *Phys. Status Solidi B: Basic Solid State Physics*, 2006, **243**, 3252–3256.
80. P. R. Lewis, R. P. Manginell, D. R. Adkins, R. J. Kottenstette, D. R. Wheeler, S. S. Sokolowski, D. E. Trudell, J. E. Byrnes, M. Okandan, J. M. Bauer, R. G. Manley and G. C. Frye-Mason, *IEEE Sensor. J.*, 2006, **6**, 784–795.
81. M. Stadermann, A. D. McBrady, B. Dick, V. R. Reid, A. Noy, R. E. Synovec and O. Bakajin, *Anal. Chem.*, 2006, **78**, 5639–5644.
82. M. Zheng, A. Jagota, E. D. Semke, B. A. Diner, R. S. McLean, S. R. Lustig, R. E. Richardson and N. G. Tassi, *Nature Mater.*, 2003, **2**, 338–342.
83. E. S. Jeng, P. W. Baroney, J. D. Nelson and M. S. Strano, *Small*, 2007, **3**, 1602–1609.
84. X. W. Tang, S. Bansaruntip, N. Nakayama, E. Yenilmez, Y. L. Chang and Q. Wang, *Nano Letters*, 2006, **6**, 1632–1636.
85. A. Star, E. Tu, J. Niemann, J. C. P. Gabriel, C. S. Joiner and C. Valcke, *Proc. Natl. Acad. Sci. USA*, 2006, **103**, 921–926.
86. E. L. Gui, L. J. Li, P. S. Lee, A. Lohani, S. G. Mhaisalkar, Q. Cao, S. J. Kang, J. A. Rogers, N. C. Tansil and Z. Q. Gao, *Appl. Phys. Lett.*, 2006, **89**, 232104.
87. K. Maehashi, K. Matsumoto, K. Kerman, Y. Takamura and E. Tamiya, *Jpn. J. Appl. Phys. Part 2-Letters & Express Letters*, 2004, **43**, L1558–L1560.
88. S. Q. Wang, E. S. Humphreys, S. Y. Chung, D. F. Delduco, S. R. Lustig, H. Wang, K. N. Parker, N. W. Rizzo, S. Subramoney, Y. M. Chiang and A. Jagota, *Nature Mater.*, 2003, **2**, 196–200.
89. F. Balavoine, P. Schultz, C. Richard, V. Mallouh, T. W. Ebbesen and C. Mioskowski, *Angew. Chem. Int. Ed.*, 1999, **38**, 1912–1915.
90. N. W. S. Kam and H. J. Dai, *J. Am. Chem. Soc.*, 2005, **127**, 6021–6026.

91. B. F. Erlanger, B. X. Chen, M. Zhu and L. Brus, *Nano Letters*, 2001, **1**, 465–467.
92. M. Shim, N. W. S. Kam, R. J. Chen, Y. M. Li and H. J. Dai, *Nano Letters*, 2002, **2**, 285–288.
93. K. Bradley, M. Briman, A. Star and G. Gruner, *Nano Letters*, 2004, **4**, 253–256.
94. A. Star, J. C. P. Gabriel, K. Bradley and G. Gruner, *Nano Letters*, 2003, **3**, 459–463.
95. M. Briman, E. Artukovic, L. Zhang, D. Chia, L. Goodglick and G. Gruner, *Small*, 2007, **3**, 758–762.
96. Y. B. Zhang, M. Kanungo, A. J. Ho, P. Freimuth, D. van der Lelie, M. Chen, S. M. Khamis, S. S. Datta, A. T. C. Johnson, J. A. Misewich and S. S. Wong, *Nano Letters*, 2007, **7**, 3086–3091.
97. K. Maehashi, T. Katsura, K. Kerman, Y. Takamura, K. Matsumoto and E. Tamiya, *Anal. Chem.*, 2007, **79**, 782–787.
98. A. D. Ellington and J. W. Szostak, *Nature*, 1990, **346**, 818–822.
99. C. Tuerk and L. Gold, *Science*, 1990, **249**, 505–510.

CHAPTER 2
Chemical and Biological Sensing Using Gold Nanoparticles

CHANG-CHENG YOU, SARIT S. AGASTI AND
VINCENT M. ROTELLO

Department of Chemistry, University of Massachusetts, 710 North Pleasant
Street, Amherst, MA 01003, USA

2.1 Introduction

Sensors play an important role in an array of areas, including biomedical
diagnosis, forensic analysis and environmental monitoring.[1] The rapid sensing
of diseases, toxic materials and bioagents will impact a wide range of quality of
life issues.

Sensors consist of a recognition element for analyte binding coupled with a
transduction element for signaling the binding event. A number of factors are
critically related to the efficiency of sensors, including response time, signal-
to-noise (S/N), sensitivity and selectivity in the presence of interfering species.
As a result, significant emphasis has been placed on improving the recognition
and transduction mechanism through the use of new materials. Significantly,
miniaturization of a sensor can increase its sensitivity.[2] To this end nanomaterials are attractive tools for sensor design due to their size, unique physicochemical properties and large surface areas.[3]

Gold nanoparticles (AuNPs) are readily fabricated through either chemical
reduction of gold salts or physical treatment of bulk gold. In addition to the
large surface areas of these materials, AuNPs feature useful optical and electronic properties coupled with low inherent toxicity.[4] The optoelectronic properties of AuNPs arise from their size, shape and local environment. The latter

Nano and Microsensors for Chemical and Biological Terrorism Surveillance
Edited by Jeffrey B.-H. Tok
© Royal Society of Chemistry, 2008
Published by the Royal Society of Chemistry, www.rsc.org

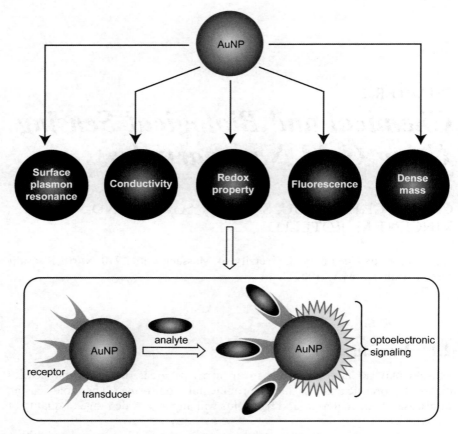

Figure 2.1   Physical properties of AuNPs and the schematic depiction of an AuNP-
              based sensor showing the optoelectronic signaling upon analyte binding.

attribute provides direct access to sensors: the binding event can alter properties
of AuNPs including surface plasmon resonance, conductivity and redox behavior (Figure 2.1). Simultaneously, AuNPs provide a versatile scaffold for functionalization with molecular or biomolecular systems, providing a versatile
platform for sensor design.[5–7]

2.2   Synthesis of Gold Nanoparticles

The synthesis of colloidal gold dates back to Faraday in 1857.[8] Recently, both
"top-down" (*e.g.* physical manipulation) and "bottom-up" (*e.g.* chemical
transformation) methods have been successfully developed to control AuNP
size, shape and surface functionalization. AuNPs can be prepared by the chemical reduction of gold salts in the presence of appropriate agents that prevent
particle coalescence (Table 2.1).

Table 2.1 Synthetic methods and capping agents for AuNPs of diverse core size.

Core size (d)	Synthetic methods	Capping agents	References
1–2 nm	Reduction of AuCl(PPh₃) with diborane or sodium borohydride	Phosphine	9–11
1.5–5 nm	Biphasic reduction of HAuCl₄ by sodium borohydride in the presence of thiol capping agents	Alkanethiol	12–14
5–8 nm	Reduction of HAuCl₄ by sodium borohydride in the presence of TOAB	Quaternary ammonium salt (TOAB)	15
8–20 nm	Reduction of HAuCl₄ by oleyl amine in water under heating	Oleyl amine	16,17
10–40 nm	Reduction of HAuCl₄ with sodium citrate in water	Citrate	18–20

Citric acid can serve as both reducing and stabilizing agent.[18,19] This protocol is usually used to prepare spherical AuNPs in diameters of 10 to 20 nm, though larger AuNPs (*e.g.* 100 nm) can also be prepared. Triphenylphosphine has been used as the capping agent to prepare small gold clusters. Highly monodisperse AuNPs in diameters of 1–2 nm are obtained by reduction of AuCl(PPh₃) with diborane[9] or sodium borohydride.[11] These small AuNPs have found interesting applications in molecular devices[10,21] and have interesting ligand exchange properties.[22]

The transfer of hydrogen tetrachloroaurate from aqueous phase to organic phases using a phase transfer agent and the subsequent reduction by sodium borohydride in the presence of alkanethiols,[12] known as Brust–Schiffrin method, leads to the relatively monodisperse AuNPs protected by thiol ligands with tunable diameters from 1.5 to 5 nm.[14] Due to the synergic effect of strong thiol–gold interactions and van der Waals attractions of the ligands, thiol-functionalized AuNPs possess excellent stability and redispersibility. Monophasic reduction of gold salts by sodium borohydride in the presence of thiols can be used to prepare water-soluble AuNPs in a single step[23,24] that provide excellent precursors for other functionalized nanoparticles.[15]

Other reducing/capping agents, such as amino acids,[25] oleyl amine,[16] as well as aliphatic and aromatic amines,[26,27] have been used to provide AuNPs. The size and shape of AuNPs can be further manipulated by conventional ripening,[28–30] digestive ripening[31,32] and UV and laser irradiation.[33,34]

The capping agents listed in Table 2.1 lack the chemical functionality required for most sensing applications. In the place-exchange process introduced by Murray *et al.*[35] the thiol ligands initially anchored on the nanoparticle surface are replaced by the presence of external thiol ligands.[36] Mixed monolayer-protected gold clusters can be obtained by place-exchange reactions using a mixture of two or more ligands as the incoming agents. The thiol ligands on gold surface possess a certain level of mobility and consequently they can undergo reposition on the surface to optimize interaction with analytes,[37] as well as slowly hop between nanoparticles.[38]

Capping agents such as citrate,[39] triphenylphosphine[40] and dimethylamino-pyridine[41] are labile and can be displaced by thiols under mild reaction conditions. As an example, citrate-stabilized AuNPs have been functionalized with oligonucleotide, protein or antibody functionalities.[42,43] Irreversible aggregation can occur during functionalization;[44] nonionic surfactants such as Tween 20 can prevent this undesired aggregation.[45]

2.3 Physical Properties of Gold Nanoparticles

Solutions of spherical AuNPs exhibit a range of colors from red/brown to violet as the core size increases from 1 to 100 nm. AuNPs generally show an intense surface plasmon band absorption peak from 500 to 550 nm[46] that arises from the collective oscillation of the conduction electrons due to the resonant excitation by the incident photons (Figure 2.2). This surface plasmon band is absent in both small nanoparticles ($d < 2$ nm) and bulk materials. Mie theory of surface plasmon resonance (SPR)[47] has been extensively correlated with the experimental results.[48,49] SPR is particularly dependent on the proximity to other nanoparticles, with the surface plasmon band red-shifting and broadening due to the interparticle plasmon coupling.[50] This phenomenon constitutes the cornerstone for their application in colorimetric sensing, which will be discussed in detail in the following sections.

AuNP concentrations can be estimated from the molar extinction coefficient of colloidal gold (520 nm, ~ 4000 M^{-1} cm^{-1} per gold atom[51]). The extinction coefficients of AuNPs with different sizes and capping ligands have been determined experimentally,[52] with a linear relationship observed between logarithms of molar extinction coefficient (ε) and core diameter (d), essentially irrespective of the ligands and solvents:

$$\ln \varepsilon = k \ln d + c \qquad (2.1)$$

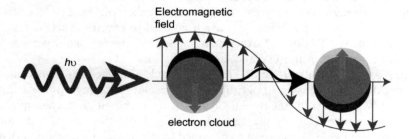

Figure 2.2   Schematic representation of the oscillation of conduction electrons across the nanoparticle in the electromagnetic field of the incident light.

where $k = 3.32$ and $c = 10.8$ ($\lambda = 506$ nm). According to Equation (2.1), an AuNP of 20 nm diameter has a molar extinction coefficient of 1×10^9 $M^{-1} cm^{-1}$, at least three orders higher than that of common organic dyes (10^4–$10^6 M^{-1} cm^{-1}$).[53]

AuNPs can photoluminesce[54,55] and can enhance fluorescence at appropriate fluorophore-to-metal distances.[56] Fluorescence resonance energy transfer (FRET) is an important deactivation pathway for the excited fluorophores in case of a good overlap between the donor's emission spectrum and the gold surface plasmon band.[57,58] Both radiative and non-radiative decay rates of fluorescent molecules are affected, resulting in highly efficient fluorescence quenching with small (1 nm) nanoparticles. Photoinduced electron transfer (PET) can also quench fluorophores in the vicinity of AuNPs,[5] and can be modulated by charging/discharging the gold core.[59]

Small metal nanoparticles display size-dependent quantization effects, leading to the discrete electron-transition energy levels. For example, 15 redox states have been observed for hexanethiol-capped AuNPs (Au_{147}, $r = 0.81$ nm) at room temperature,[60] indicating that AuNPs feature molecule-like redox properties.[61] The quantized capacitance charging behavior of AuNPs is distinctly affected by electrolyte ions, external ligands and applied magnetic field.[21]

2.4 Colorimetric Sensing

Clustering of AuNPs ($d > 3.5$ nm) results in interparticle surface plasmon coupling,[62] with a concomitant color change from red to blue. These color changes can be readily observed by the naked eye at nanomolar concentrations, making them excellent candidates for colorimetric sensing.[3]

Colorimetric detection of metal ions can be achieved through the incorporation of chelating agents onto the nanoparticle surface (Figure 2.3). AuNPs

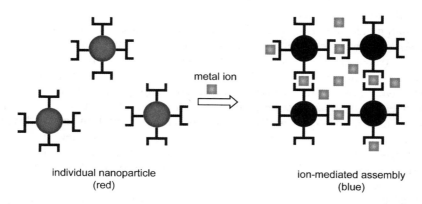

individual nanoparticle
(red)

ion-mediated assembly
(blue)

Figure 2.3 Schematic illustration of metal ion-induced nanoparticle aggregation.

carrying 15-crown-5 moieties have been used to detect potassium ions from µM to mM concentrations in the presence of high concentrations of Na^+.[63] Sensing of Na^+ has been obtained through the incorporation of 12-crown-4 onto AuNPs.[64] Similarly, phenanthroline-functionalized and lactose-functionalized AuNPs have been fabricated to detect Li^+ and Ca^{2+} respectively with little interference from other metal ions.[65,66]

Hupp *et al.* have developed a heavy metal ion sensor using 11-mercaptoundecanoic acid (MUA)-functionalized AuNPs,[67] with the surface carboxylates acting as metal ion receptors driving aggregation. The presence of Pb^{2+}, Cd^{2+} or Hg^{2+} ($\geq 400\,µM$) induces aggregation, which can be modulated through buffer choice.[68] Mixed monolayer-protected AuNPs with carboxylate and 15-crown-5 functionalities provide a colorimetric sensor for Pb^{2+} where the ions disrupt the initially hydrogen-bonded assemblies.[69] Cysteine and peptide-functionalized AuNPs have been used to detect Cu^{2+} and Hg^{2+}, respectively[70,71] at sub-micromolar concentrations. Recently, a detection limit of 100 nM has been reported for Hg^{2+} using DNA-functionalized AuNPs.[72] Liu and Lu have devised a highly selective lead sensor exploiting DNAzyme (catalytically active DNA molecules) to control assembly of AuNPs[73-75] that show metal-dependent activities. The DNAzyme and the DNA-functionalized AuNPs initially form blue assemblies through Watson–Crick base pairing.[76,77] In the presence of Pb^{2+} (100 nM), the DNAzyme is activated, cleaving the substrate strand to dissemble the AuNPs, resulting in a color change to red. The sensor is highly specific, with no response observed with other divalent metal ions such as Ca^{2+}, Co^{2+}, Ni^{2+} and Cd^{2+}. Alignment of AuNPs in the assemblies plays an important role in the disassembly process,[78] with optimization of the AuNP orientation allowing rapid detection at ambient temperature.[75,78] This approach has been used for the detection of adenosine using AuNPs functionalized with an adenosine-specific aptazyme.[79]

In the area of anion sensing, Kubo *et al.* have attached isothiouronium groups onto AuNPs for colorimetric sensing of oxoanions.[80] Oxoanions such as AcO^-, HPO_4^{2-} and malonate induce color changes, whereas other anions (*e.g.* Cl^-) are non-interfering. Thioglucose-coated AuNPs have been used to sense fluoride anions in water.[81]

Small molecule sensing is challenging, as these species are generally not able to serve as bridging agents for nanoparticle assembly. Geddes *et al.* have demonstrated glucose sensing by using assemblies of concanavalin A (Con A) and dextran-functionalized AuNPs.[82,83] Con A is a multivalent protein (four sugar binding sites) that cause dextran-coated nanoparticles to cross-link. In the presence of glucose, competitive interactions with Con A release the individual dextran-coated AuNPs, which can be monitored by either UV/Vis spectrometry[82] or wavelength-ratiometric resonance light scattering.[83] A glucose dynamic sensing range of 1–40 mM was achieved, well suited for diagnostic applications as the blood glucose level in healthy people is 3–8 mM and in diabetics 2–40 mM.

Incorporation of AuNPs into molecularly imprinted polymers (MIPs) provides SPR-based molecular sensors. Immobilization of MUA-AuNPs into a

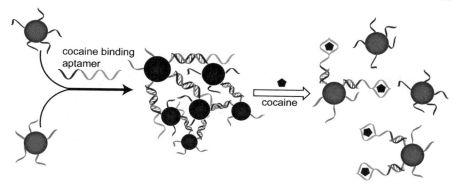

Figure 2.4 Schematic representation of the colorimetric detection of cocaine based on the target-induced disassembly of nanoparticle aggregates linked by corresponding aptamers.

MIP gel placed between two glass slides provided a colorimetric adrenaline sensor with a dynamic range from 5 μM to 2 mM.[84]

Aptamers are single-stranded oligonucleic acids that can bind targets with high affinity and specificity.[85] Lu and co-workers have designed an effective cocaine sensor using a cocaine aptamer-mediated AuNP assembly (Figure 2.4).[86] When cocaine (50 to 500 μM) is present in the system, the aptamer changes its structure to bind cocaine, resulting in the disassembly of the aggregates with a concomitant blue-to-red color change. This approach is general, as any aptamer of choice could be engrafted into this system,[86] allowing multiplex sensing.[87] Both adenosine and cocaine aptamer-linked nanoparticle aggregates have been incorporated into a lateral flow device, allowing a "dipstick" test that can be performed on human blood serum.[88]

DNA-mediated AuNP assembly was demonstrated by Mirkin *et al.* in 1996,[76] and has been extensively used in the colorimetric detection of oligonucleotides[77,89–96] and fabrication of structured assemblies.[97] The general approach for the detection of oligonucleotides is to use two ssDNA-modified AuNPs where the base sequences are complementary to the ends of the target oligonucleotides. The intense absorptivity of AuNPs coupled with the strong and highly specific base-pairing of DNA molecules enables the ultrasensitive detection of oligonucleotides in a quantitative manner. When large AuNPs (*e.g.* 50 nm or 100 nm) are employed, the detection limit was subpicomolar (PCR).[90] Interestingly, citrate-stabilized AuNPs can discern ssDNA and dsDNA at the level of 100 fmol based on simple electrostatic interactions.[98] Oligonucleotide-directed AuNP assembly can be used for the colorimetric screening of DNA binders[99] and triplex DNA binders (Figure 2.5).[100] The simplicity of this approach makes it convenient for the screening of potential triplex binders from large combinatorial libraries.[100]

Ligand-modified AuNPs provide a useful platform for the colorimetric detection of proteins. An example is agglutinin, a bivalent lectin that specifically recognizes β-D-galactose (1 ppm), inducing the aggregation of galactose-functionalized AuNPs.[101] Protein-directed glyconanoparticle assembly has

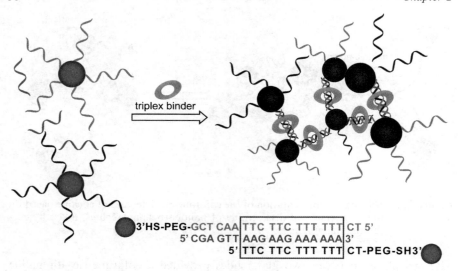

Figure 2.5 Schematic depiction for the screening of DNA triplex binder using DNA-directed AuNP assembly.

likewise been used to detect protein–protein interactions. In one study, Chen *et al.* used assemblies Con A and mannose-modified AuNPs to identify and quantify binding partners for Con A.[102] Recently, the controlled aggregation of glyconanoparticles has been harnessed to attain colorimetric detection of cholera toxin.[103] In fundamental studies, biotin-functionalized AuNPs have been deposited on glass substrates with colorimetric changes upon streptavidin binding demonstrating nanoparticle size-dependent sensitivity.[104,105]

AuNPs carrying an aptamer specific to platelet-derived growth factors (PDGFs) have been used to detect PDGFs and identify PDGF receptors.[106] A straightforward aptamer-based colorimetric sensing system for thrombin was reported by Dong *et al.*[107] The thrombin-binding aptamers fold into a structure of G-quadruplex/duplex in the presence of thrombin due to the aptamer-protein recognition, resulting in aggregation. Willner *et al.* used thrombin aptamer-modified glass surface for sensing.[108] Treatment of the functionalized glass substrate thrombin targets and aptamer-functionalized AuNPs provided a "sandwich" complex. The immobilized AuNPs were grown using HAuCl$_4$, CTAB and NADH.[109]

Dithiols have long been used as cross-linkers in the assembly of AuNPs.[110] Dithiol-functionalized peptides provide a sensitive platform for colorimetric detection of proteases at the low nanomolar range;[111] Scrimin and co-workers designed cysteinyl derivatives of peptide substrates specific to thrombin and lethal factor.[112] The peptides were treated with analytes and then incubated with citrate-stabilized AuNPs. The intact peptides cause nanoparticle aggregation, whereas the protease-cleaved peptides do not. Stevens *et al.* simplified this approach using AuNPs functionalized with Fmoc-protected peptides (substrate of thermolysin) with a cysteine anchor. The π–π stacking interactions between Fmoc groups lead to an assembled state of AuNPs, which is disrupted through peptide digestion by thermolysin. This system has impressive sensitivity of 90 zg mL^{-1} (fewer than 380 molecules).[113] Enzymatic cleavage of DNA molecules provides a screening method for endonuclease activity and inhibition.[114] Similarly, enzyme-triggered AuNP assembly/disassembly has been employed to detect kinase[115] and phosphatases.[116,117] AuNPs can also act as a colorimetric sensor for protein conformational changes.[118]

2.5 Fluorescence Sensing

The exceptional quenching ability of AuNPs makes them excellent energy acceptors for the FRET assays.[57] Poly-pyridyl complex $[Ru(bpy)_3]^{3+}$, for example, is effectively quenched by anionic tiopronin-coated AuNPs.[119] These electrostatic interaction-driven complexes can be dissociated by addition of electrolytes such as K^+, Bu_4N^+ and Ca^{2+} salts.[119] Rhodamine B-adsorbed AuNPs have been used in the turn-on fluorescence sensing of Hg^{2+},[120] with the selectivity of this system for Hg^{2+} 50-fold greater than other divalent metal ions (*e.g.* Pb^{2+}, Cd^{2+}, Co^{2+}) with a detection limit of 2.0 ppb. Nile red-adsorbed AuNPs show selective detection of aminothiols,[121] while Zhu, Li and collaborators have fabricated bispyridyl perylene-bridged AuNPs as Cu^{2+} sensors.[122] More recently, lanthanum complexes of bipyridine-functionalized AuNPs have been used as phosphorescent sensors for alkali earth and transition metal ions.[123]

AuNP-based molecular beacons have been used to sense DNA (Figure 2.6).[124] This AuNP-based molecular beacon exhibits a sensitivity enhanced up to 100-fold relative to molecular quenchers. Nie *et al.* have likewise shown that oligonucleotide-functionalized AuNPs spontaneously assemble into a constrained arch-like conformation with close donor-acceptor distance. This structure responds to target ssDNA through a hybridization-induced strand stretching with concomitant restoration of fluorescence.[125] Distance-dependent FRET likewise provides a potent means for the detection of DNA cleavage.[126]

The emission of semiconductor quantum dots (QDs) is quenched by AuNPs. Melvin *et al.* have designed a fluorescent competitive assay for DNA detection

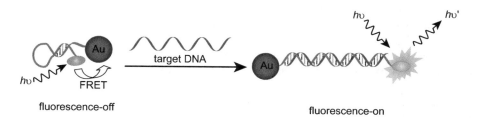

Figure 2.6 Schematic drawings of the conformational changes of the dye-oligonucleotide-AuNP conjugates before and after hybridization with the target DNA.

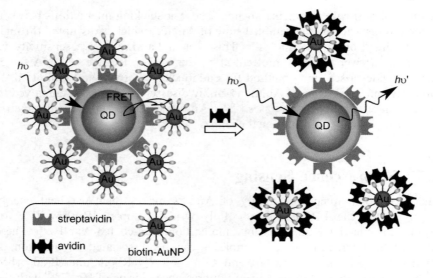

Figure 2.7   Competitive inhibition assay for the detection of avidin by using QD-AuNP couple.

using QDs and AuNPs as the FRET donor-acceptor couple.[127] Kim and co-workers have devised an inhibition assay of proteins on the basis of FRET between QDs and AuNPs (Figure 2.7).[128] The streptavidin-functionalized QDs in this approach specifically interact with biotinylated AuNPs to provide quenched assemblies *via* a QD-to-AuNP FRET process. In an analogous process, the assembly of dextran-functionalized QDs and Con A-coated AuNPs has been used to detect glycoproteins.[129]

In an alternative approach to specific protein sensors, Rotello and collaborators have coupled an indicator-displacement assay with a "chemical nose" approach to fabricate a protein sensor array. This array was composed of six cationic AuNPs and one anionic poly(*p*-phenyleneethynylene) (PPE) polymer (Figure 2.8a).[130] The initially quenched polymer-AuNP complexes are displaced by proteins, resulting in fluorescence restoration. Because the protein-nanoparticle interactions are determined by their respective structural features such as charge, hydrophobicity and hydrogen bonding,[131,132] the differing particle-protein affinities generate a distinct fluorescence response pattern for individual proteins (Figure 2.8b). Linear discrimination analysis (LDA) was then used to differentiate the response patterns with high accuracy. Protein samples of unknown identity and concentration were identified with 94.2% accuracy while the protein concentrations were generally determined within ±5% error.

This array-based sensing based on AuNP-conjugated polymer systems has been extended to the rapid and convenient detection of pathogens,[133] an issue of great medical and environmental importance. In these studies three cationic AuNPs and one anionic PPE with their pendent carboxylate and oligo(ethylene

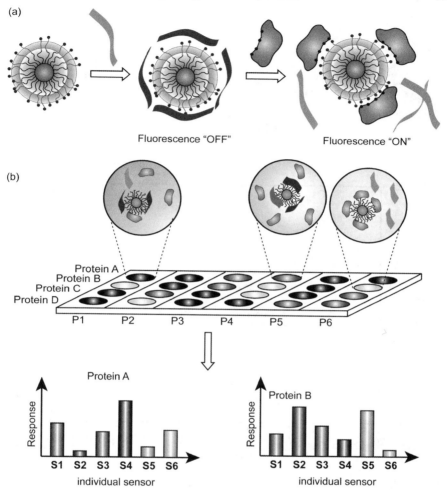

Figure 2.8 Schematic illustration of a "chemical nose" sensor array based on AuNP-fluorescent polymer conjugates. (a) The competitive binding between protein and quenched polymer-AuNP complexes leads to the fluorescence light-up. (b) The combination of an array of sensors generates fingerprint response patterns for individual proteins.

glycol) residues provided excellent differentiation of bacteria. The initially quenched PPE polymers generate a fluorescence recovery pattern in the presence of bacteria relying on the diverse binding strength of polymer-nanoparticle and bacteria-nanoparticle interactions. Twelve microorganisms were identified using this sensor array. The list of identified microorganisms contains a wide variety of Gram-positive (*e.g. A. azurea, B. subtilis*) as well as Gram-negative (*e.g. E. coli, P. putida*) species. As shown in Figure 2.9, this sensor array successfully discerns not only the species, but also the strains of the bacteria by using LDA analysis of

(a)

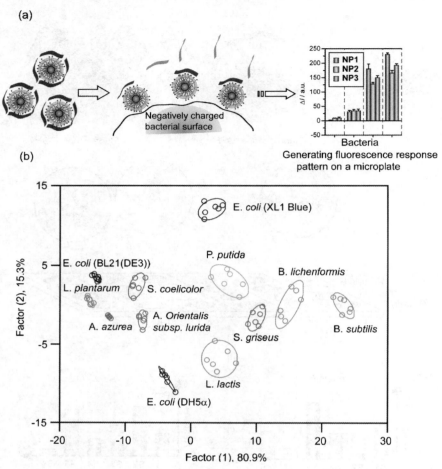

(b)

Figure 2.9 (a) Schematic representation of the generation of fluorescence response patterns from negatively charged bacterial surfaces. (b) Canonical score plot for the fluorescence response patterns of three AuNP-conjugated polymer constructs in the presence of bacteria processed with LDA. The first two factors consist of 96.2% variance and the 95% confidence ellipses for the individual bacteria are depicted.

the fluorescence response patterns. The exceptional quenching ability of AuNPs and the "molecular wire" effect of PPE polymer are the both key to the performance of this system.[134]

2.6   Surface-enhanced Raman Scattering-based Sensing

Raman scattering is sensitive to different vibrational modes, providing a fingerprint of the target molecules. The cross section for Raman scattering is 10 to 15 orders of magnitude smaller than that of fluorescence, limiting the direct application of this technique. When a molecule is adsorbed on rough metal surfaces, however, its Raman scattering can be enhanced by up to 15 orders of magnitude, enabling single molecule detection.[135] Surface-enhanced Raman scattering (SERS) is attributed to the local electromagnetic-field enhancement induced by plasmon resonance of nanoparticles,[136,137] with the field enhancement dependent on size, shape, orientation and aggregation of the nanomaterial.

AuNP-based SERS has been employed extensively in biological sensing. Mirkin *et al.* used AuNP probes labeled with oligonucleotides and Raman-active dyes for multiplexed detection of oligonucleotide targets with a detection limit of 20 fM (Figure 2.10).[138] This method can discriminate single nucleotide polymorphisms present in six different viruses. Non-fluorescent Raman tags have been incorporated into DNA-functionalized AuNP probes for SERS detection of DNA,[139] with simultaneous identification of up to eight probes in a mixture and a detection limit of *ca.* 100 nM.

AuNPs functionalized with Raman dyes and either protein ligands or antibodies have been used to detect the protein–small molecule and protein–protein interactions.[140] Lipert, Porter and co-workers have functionalized AuNPs with a monolayer of 5-thiol-2-nitrobenzoate (a strong Raman scatterer) followed by covalently linked antibodies for SERS detection.[141] This immunoassay system

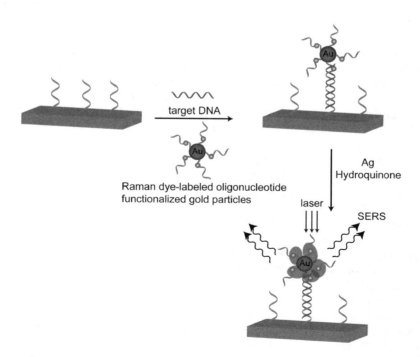

Figure 2.10 AuNP probes labeled with oligonucleotides and Raman-active dyes for multiplexed detection of oligonucleotide targets.

was used for the detection of free prostate-specific antigen (PSA) with a detection limit of $\sim 1\,pg\,mL^{-1}$ in human serum through a sandwich assay format employing on monoclonal antibodies.

2.7 Electrical and Electrochemical Sensing

The roughening of the conductive sensing interface, the catalytic properties and the conductivity properties of AuNPs make them excellent candidates for electroanalytical applications.[142] Multilayered AuNPs on electrode surfaces enhance surface area, providing enhanced electrochemical detection of redox analytes.[143] AuNP-functionalized electrodes feature high electrocatalytic activity compared with bulk electrodes, allowing them to discriminate the voltammetric signals of dopamine and ascorbate.[144]

Chemiresistors are solid-state devices that change electrical resistance changes after interaction with chemical species, and can be tuned through particle size, interparticle separation, ligand properties and chemical environments.[145] These materials have been widely used for vapor sensing.[142] Wohltjen and Snow have created a sensor through deposition of a thin film of octanethiol-coated AuNPs ($d \sim 2$nm) onto an interdigitated electrode that can detect toluene and tetrachloroethylene, with a detection limit of *ca.* 1 ppm.[146] AuNPs bearing polar functional groups provide chemiresistors that are more sensitive to polar analytes.[147]

Vossmeyer *et al.* have used films formed by dodecylamine-stabilized AuNPs and α,ω-dithiols with different chain lengths (C_6, C_9, C_{12}, C_{16})[148] for vapor sensing, observing that the resistance at a given concentration of toluene analyte increases exponentially with increasing number of methylene units in the monolayer. Zhong and co-workers have explored the performance of sensor arrays constructed from AuNP films and interdigitated microelectrodes,[149,150] establishing the correlation between the vapor-response sensitivity and the interparticle spacing.

Layer-by-layer AuNP-dendrimer self-assemblies composites have also been used in vapor sensing.[151] The chemical selectivity of the films was determined by the polarity of the dendrimer. More recently, a bioconjugate material was made by reduction of gold salts onto spider silk.[152] This material was able to determine the polarity of alcohol vapors (from methanol to butanol) through conductivity changes.

The electroactivity of AuNPs coupled with the complexation features of macrocyclic compounds provides useful sensor systems. Willner and co-workers have created three-dimensional electrode surfaces, with AuNPs and oligocationic cyclophanes or molecular squares alternately deposited onto a functionalized indium-doped tin oxide (ITO) conductive glass (Figure 2.11).[153–156] The AuNPs provide conductive surfaces, while the macrocycles serve as "molecular glues" that bind π-donor substrates such as hydroquinone. Binding enhances the local concentration of substrates at the electrode surface, enhancing electrode sensitivity,[157] an effect that can be readily tuned by adjusting the number

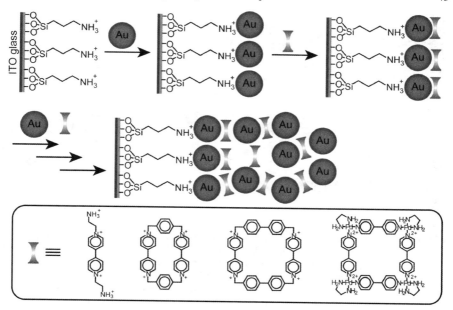

Figure 2.11   Electroactive multilayers formed by the self-assembly of anionic AuNPs and bipyridinium-based oligocations.

of the assembled layers.[153] The binding affinity between the macrocycles and the analytes determines the concentration of substrates at the electrode, providing selectivity.[155] Sensing studies on anionic π-donor analytes confirm that the preconcentration of electroactive species at the electrode arises from specific π donor-acceptor interactions.[154]

Willner *et al.* have deposited a film consisting of polyethyleneimine, AuNPs and cyclobis(paraquat-*p*-phenylene) on the Al_2O_3 insulating layer of an ion-sensitive field-effect transistor to allow the analysis of non-redox-active compounds.[158] It was demonstrated with adrenaline as a model compound that either the source-drain current or the gate-source voltage could be used for analysis. Recently, AuNP-based microelectronic devices were used to detect H_2S, with adsorption of H_2S molecules onto the nanoparticle modulating the hopping behavior of electrons through the particles.[159]

While bulk Au metal is relatively inert, AuNPs have impressive catalytic activity due to their large surface-to-volume ratio and interface-dominated properties.[4,160] AuNP-modified electrodes have been employed for the electrocatalytic detection of organic and biological molecules.[144,161–163] Jena and Raj have demonstrated the electrocatalytic sensing of glucose using an AuNP-coated gold electrode,[162] with a detection limit of 50 nM.

The direct immobilization of redox enzymes onto the electrodes generally results in low sensitivity due to poor electrical communication between the enzyme and the electrode. Co-deposition of AuNPs and redox enzymes can

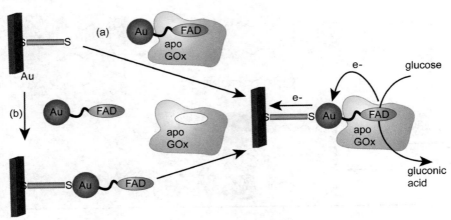

Figure 2.12 Fabrication of GOx electrode by the reconstitution of apo-GOx on a
FAD-functionalized AuNP. (a) The adsorption of AuNP-reconstituted
apo-GOx to a dithiol monolayer assembled on a gold electrode and
(b) the adsorption of FAD-AuNPs on the dithiol-covered gold electrode
followed by the reconstitution of apo-GOx on the functional nanoparticles. Both routes afforded almost identical monolayers.

improve this electrical communication.[164–167] Additionally, immobilization of
enzymes onto AuNPs can increase turnover rates,[168,169] enhancing sensor
sensitivity. Willner and co-workers have constructed a bioelectrocatalytic system through the reconstitution of apo-glucose oxidase (apo-GOx) on a 1.4-nm
gold nanoparticle that was functionalized with the cofactor flavin adenine
dinucleotide (FAD) (Figure 2.12).[170] The electrodes exhibited very efficient
electrical communication with the electrode, with electron-transfer turnover
rates seven-fold higher than the electron-transfer rate constant of native GOx.

AuNP-based electrochemical detection of DNA provides an alternative to
optical sensors.[171] Mirkin *et al.* have developed a DNA array detection method
where the binding of oligonucleotide-functionalized AuNPs causes conductivity
changes (Figure 2.13).[172] Short capture oligonucleotides were immobilized on
the SiO_2 surface between two electrodes. Target DNA and AuNPs functionalized with oligonucleotides were then added and the enlargement of metal
nanoparticles was performed by deposition of silver onto AuNPs. Target DNA
has been detected using this method at concentrations as low as 500 femtomolar
with a point mutation selectivity factor of $\sim 100000 : 1.$[172]

The redox properties of AuNPs enable their use as electrochemical labels for
oligonucleotide detection. Ozsoz *et al.* have shown that the interaction of a
target DNA-modified electrode with complementary AuNP-conjugated probes
generated a gold oxide wave at $+1.2$ V.[173] Detection limits of this system were
0.78 fmol with the assistance of PCR amplification. Other readily accessible
amplification tactics, including silver deposition[174] and the incorporation of
electrochemically active groups onto AuNPs,[175,176] have also been explored. As
an example, ferrocene-capped AuNP/streptavidin conjugates were conjugated

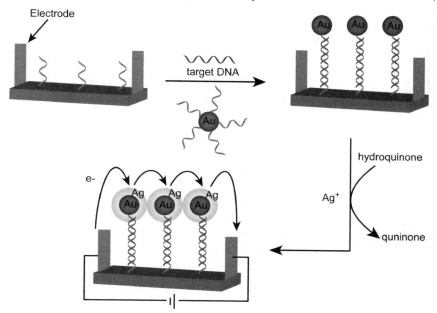

Figure 2.13 Electrical detection of DNA based on the "sandwich" hybridization with DNA-functionalized AuNPs followed by silver deposition.

to a biotinylated DNA detection probe of a "sandwich" DNA complex on the electrode, providing a detection limit of 2 pM.[175] Fan *et al.* have recently developed a sandwich detection system featuring the hybridization with AuNP-labeled reporter probe DNA followed by treatment with $[Ru(NH_3)_6]^{3+}$ complexes.[177] Willner *et al.* reported a novel amplified electrochemical detection of DNA through the aggregation of AuNPs on electrodes and the intercalation of methylene blue into the DNA-cross-linked structure,[178] with a detection limit of 0.1 pM.

AuNP-based immunoassays using either conductivity changes or electrochemical signaling have been used for protein sensing. A conductivity immunoassay developed by Velev and Kaler used the adsorption of proteins between antibodies immobilized in an electrode gap with a secondary AuNP-tagged antibody followed by the enlargement of AuNPs,[179] detecting human IgG at 0.2 pM. Limoges *et al.* devised an electrochemical immunoassay using AuNP-labeled antibodies (Figure 2.14),[180] where the AuNP-labeled antibody forms sandwich complexes with goat IgG target and the immobilized antibody. The AuNPs were dissolved in an acidic bromine–bromide solution to release gold ions which are quantitatively determined at a disposable carbon-based screen-printed electrode (SPE) using anodic stripping voltammetry (ASV) with a detection limit of 3 pM. Another strategy based on cyclic accumulation of AuNPs has been developed for determination of human IgG by ASV,[181] where the probe antibody in the sandwich complexes is labeled with dethiobiotin and

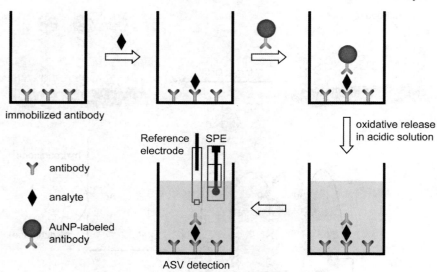

immobilized antibody

Reference | SPE
electrode

oxidative release
in acidic solution

Y   antibody

◆   analyte

AuNP-labeled
antibody

ASV detection

Figure 2.14   Schematic representation of the sandwich electrochemical immunoassay
with an AuNP label.

avidin-AuNPs are introduced for further complexation. The alternating treatment of the system with biotin and avidin-AuNPs enhanced the signal, providing detection limits of $0.1\,ng\,mL^{-1}$ human IgG.

2.8 Gold Nanoparticle Amplified SPR Sensing

SPR is sensitive to the refractive index of layers present in the interfacial region,
providing access to the sensing system.[182] The introduction of AuNPs onto
sensor surfaces amplifies this effect by the high mass and dielectric constant of
AuNPs and the electromagnetic coupling between AuNPs and the metal
film.[183] As an example, a MIP gel with embedded AuNPs has been prepared on
a gold film-coated chip,[184] with analyte binding increasing the distance between
the AuNPs and the substrate surface. This enhanced approach lowered the
limit of detection of dopamine to nanomolar concentrations,[184] as compared to
the micromolar sensitivity using a colorimetric method.[84]

The sensitivity of oligonucleotide detection can be drastically improved by
using AuNP-amplified SPR.[185,186] Keating *et al.* designed a "sandwich"
approach with a monolayer of 12-mer oligonucleotides immobilized onto a
gold substrate and the target DNA and AuNPs carrying complementary DNA
molecules were combined successively.[185] The AuNP-tagged surface provides
a greater than 10-fold increase in angle shift, corresponding to a more than
1000-fold improvement in sensitivity and, a $\sim 10\,pM$ limit of quantification. A
dextran layer between the gold film and the immobilized DNA molecules
effectively reduces the non-specific adsorption, leading to detection of a 39-mer

DNA at femtomolar level.[187] Real-time DNA detection has been realized by using ssDNA-modified AuNPs and micropatterned chemoresponsive diffraction gratings.[188]

Antibody-antigen recognition allows specific protein sensing.[189] Natan *et al.* devised the first AuNP-enhanced SPR immunosensing system using either antigen- or antibody-functionalized AuNPs as signal amplifiers.[190] Picomolar detection of h-IgG was realized using such particle enhancement. Recently, both competitive and sandwich immunoassays have been developed to quantify inhibition of metalloproteinases-2 using SPR.[191]

2.9 Quartz Crystal Microbalance-based Sensing

A quartz crystal microbalance (QCM) measures mass changes on a quartz crystal resonator through changes in frequency upon substrate binding.[192] AuNPs and AuNP-dendrimer acting as sorptive materials have been used for vapor sensing.[193–195] AuNPs have also been used as "mass enhancers" providing a powerful approach for promoting detection sensitivity by amplifying frequency changes.[196]

Efforts have been focused on the detection of oligonucleotides using QCM sensors with the assistance of AuNPs. Introduction of a layer of AuNPs between the gold film and the immobilized ssDNA significantly improves the detection capacity of DNA sensing as a consequence of the large surface of AuNPs.[197] "Sandwich" approaches can likewise improve the detection limit when one end of the target oligonucleotides hybridizes with the immobilized ssDNA molecules (recognition elements) and the other end hybridizes with ssDNA-modified AuNPs.[198–202] Multivalent ssDNA-modified AuNPs can provide dendritic amplification.[199] Catalyzed deposition of gold onto the amplifier AuNPs has also been used to improve the sensitivity of QCM detection of DNA, with a detection limit of *ca.* 1 fM attained.[203,204] Recently, a microcantilever-based DNA sensor has been created where the sensor element is 100 times smaller than a QCM element, enabling the construction of a high-density senor array for multiplexed detection.[205] Using a "sandwich" approach coupled with AuNP-amplification, a detection limit of 23 pM was achieved on a microcantilever for sensing a 30-mer DNA.[205]

Detection of streptavidin on a QCM using AuNPs as signal amplification probes has been reported,[206] using biotinylated BSA immobilized on the gold surface of the QCM electrode. Treatment of the resulting interface with biotin-functionalized AuNPs enhanced the frequency change of two-fold providing a dynamic range of $1 \, ng \, mL^{-1}$ to $10 \, \mu g \, mL^{-1}$.

2.10 Gold Nanoparticle-based Bio-barcode Assay

Mirkin's group has demonstrated highly multiplexed and sensitive detection of proteins and nucleic acids using an AuNP-based bio-barcode assay.

The bio-barcode assay was first employed to analyze PSA.[207] A magnetic microparticle carrying antibodies that specifically bind PSA was first used to capture the protein (Figure 2.15a). Subsequently, a AuNP functionalized with barcode DNA and antibodies unique to the protein target was combined to sandwich the protein between magnetic and gold particles. Magnetic separation followed by thermal dehybridization of DNA on the AuNPs yielded the ssDNA-functionalized AuNPs and free barcode nucleic acids that was directly analyzed by sandwich hybridization with ss-DNA functionalized AuNP probes followed by silver amplification, leading to a detection limit of 30 aM. PCR amplification on the barcode DNA provided an unprecedented sensitivity of 3 attomolar.[207] This bio-barcode approach has also been used to detect amyloid-β-derived diffusible ligand, a soluble pathogenic Alzheimer's disease marker in cerebrospinal fluid.[208] This approach has been used to achieve multiplexed detection of protein cancer markers.[209]

To simplify signal readout, fluorophore-tagged bio-barcoded AuNPs have been used in the protein detection.[210] With PSA, a detection limit of 300 aM was obtained. Nam and Groves *et al.* developed a colorimetric bio-barcode assay for the protein detection where the released barcode nucleotides serve as a bridging agent of two ssDNA-functionalized barcode capture AuNPs.[211,212] Through this method, cytokine has been detected at 30 aM.

The ultrasensitive detection of DNA has also been obtained using bio-barcode amplification.[213,214] The antibodies in the protein detection system are replaced by specific ssDNA (Figure 2.15b). The target DNA can hybridize with both the magnetic particle probes and the bio-barcoded AuNP probes to form sandwich assemblies. Magnetic separation followed by thermal dehybridization releases the free barcode nucleotides for analysis, providing 500-zeptomolar sensitivity.[213] Multiplexed DNA detection is likewise possible by using a mixture of different bio-barcoded nanoparticle probes.[214]

2.11 Conclusions

AuNPs provide an adaptable platform for the incorporation of an enormous array of functionality ranging from small organic ligands to large biomacromolecules. These materials also feature physical, optical and electronic attributes arising from the particle core. Taken together, functionalized AuNPs can serve as both molecular receptor and signal transducer for sensing processes.

The sensitivity of AuNP-based sensors is determined by the analyte, recognition partner and the transduction mechanism. The detection limit of AuNP sensors ranges from micromolar to zeptomolar, depending on the target species and the design of the sensor (Table 2.2). These receptors are often near the theoretical detection limit for practical time scales due to the analyte transport limitation,[2] making directed transport of biomolecules imperative for further enhancements in sensitivity.

Another important issue is modulation of the nanoparticle surface functionality for selective capture of target analytes. Highly selective double-stranded

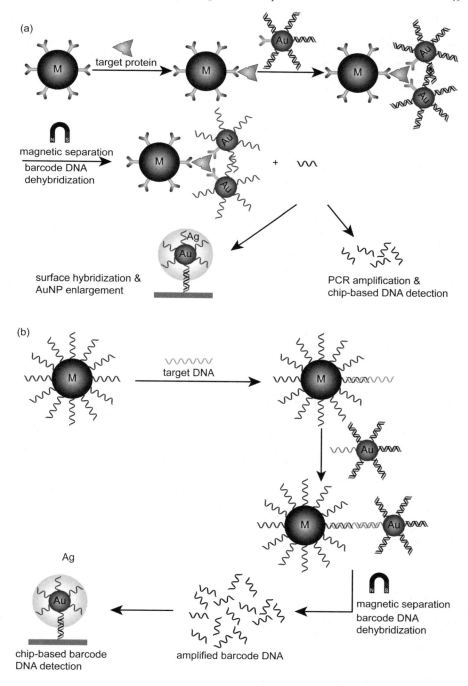

Figure 2.15 AuNP-based bio-barcode assay of (a) proteins and (b) DNA.

Table 2.2 Comparison of detection limits of different AuNP-based chemical and biological sensors.

Assay method	Analytes	Detection limit	Reference
Colorimetric	Metal ion (Pb^{2+})	2.5 nM	69
	Organic compound (adrenaline)	5 µM	84
	Oligonucleotide (ssDNA 30 mer)	50 pM	90
	Protein (thermolysin)	$\sim$ 1 aM	113
Fluorescence	Metal ion (Hg^{2+})	10 nM	120
	Oligonucleotide (ssDNA 16 mer)	0.2 µM	124
	Protein (avidin)	10 nM	128
Electrochemical	Volatile compound (toluene)	1 ppm	146
	Organic compound (glucose)	50 nM	162
	Oligonucleotide (ssDNA 27 mer)	500 fM	172
	Protein (human IgG)	0.2 pM	179
SERS	Oligonucleotide (ssDNA 30 mer)	20 fM	138
	Protein (PSA)	30 fM	141
SPR	Organic compound (dopamine)	1 nM	184
	Oligonucleotide (ssDNA 39 mer)	1.38 fM	187
	Protein (human IgG)	6.7 pM	190
QCM	Volatile compound (toluene)	10 ppm	193
	Oligonucleotide (ssDNA 27 mer)	$\sim$ 1 fM	203
	Protein (streptavidin)	$\sim$ 20 pM	206
Bio-barcode	Oligonucleotide (ssDNA 27 mer)	500 zM	213
	Protein (PSA)	30 aM	207

DNA, antibody–antigen and aptamer–analyte interactions have been used for recognition. These systems present challenges in the high-throughput screening of different analytes due to the number of recognition elements required for multi-analyte detection. In this regard, a sensor array approach could play an important role as selectivity is required rather than specificity, allowing a limited number of individual sensors to provide nearly unlimited screening capability.

Overall, AuNPs provide highly versatile and tractable scaffolds for the creation of sensing systems. Given their ease of production, stability and ease of functionalization, the applications for sensor systems based on these materials are limited only by the imaginations of the research community.

Acronyms

apo-GOx	Apo-glucose oxidase
ASV	Anodic stripping voltammetry
AuNPs	Gold nanoparticles
Con A	Concanavalin A
dsDNA	Double stranded DNA
FAD	Flavin adenine dinucleotide
FRET	Fluorescence resonance energy transfer

ITO   Indium-doped tin oxide
LDA   Linear discrimination analysis
MIPs   Molecularly imprinted polymers
MUA   11-mercaptoundecanoic acid
PCR   Polymerase Chain Reaction
PDGFs   Platelet-derived growth factors
PET   Photoinduced electron transfer
PPE   poly(*p*-phenyleneethynylene)
PSA   Prostate-specific antigen
QCM   Quartz crystal microbalance
QDs   Quantum dots
SERS   Surface enhanced Raman scattering
SPE   Screen-printed electrode
SPR   Surface plasmon resonance
ssDNA   Single stranded DNA

References

1. D. Diamond, in *Principles of Chemical and Biological Sensors*, D. Diamond ed., John Wiley & Sons, Inc., 1998, pp. 1–18.
2. P. E. Sheehan and L. J. Whitman, *Nano Lett.*, 2005, **5**, 803–807.
3. N. Rosi and C. A. Mirkin, *Chem. Rev.*, 2005, **105**, 1547–1562.
4. M.-C. Daniel and D. Astruc, *Chem. Rev.*, 2004, **104**, 293–346.
5. K. G. Thomas and P. V. Kamat, *Acc. Chem. Res.*, 2003, **36**, 888–898.
6. E. Katz and I. Willner, *Angew. Chem. Int. Ed.*, 2004, **43**, 6042–6108.
7. C. C. You, A. Verma and V. M. Rotello, *Soft Matter*, 2006, **2**, 190–204.
8. M. Faraday, *Philos. Trans.*, 1857, **147**, 145–181.
9. G. Schmid, R. Pfeil, R. Boese, F. Bandermann, S. Meyer, G. H. M. Calis and W. A. Vandervelden, *Chem. Ber.*, 1981, **114**, 3634–3642.
10. G. Schmid, *Chem. Rev.*, 1992, **92**, 1709–1727.
11. W. W. Weare, S. M. Reed, M. G. Warner and J. E. Hutchison, *J. Am. Chem. Soc.*, 2000, **122**, 12890–12891.
12. M. Brust, M. Walker, D. Bethell, D. J. Schiffrin and R. Whyman, *J. Chem. Soc., Chem. Commun.*, 1994, 801–802.
13. D. V. Leff, P. C. Ohara, J. R. Heath and W. M. Gelbart, *J. Phys. Chem.*, 1995, **99**, 7036–7041.
14. M. J. Hostetler, J. E. Wingate, C.-J. Zhong, J. E. Harris, R. W. Vachet, M. R. Clark, J. D. Londono, S. J. Green, J. J. Stokes, G. D. Wignall, G. L. Glish, M. D. Porter, N. D. Evans and R. W. Murray, *Langmuir*, 1998, **14**, 17–30.
15. J. Fink, C. J. Kiely, D. Bethell and D. J. Schiffrin, *Chem. Mater.*, 1998, **10**, 922–926.
16. M. Aslam, L. Fu, M. Su, K. Vijayamohanan and V. P. Dravid, *J. Mater. Chem.*, 2004, **14**, 1795–1797.
17. H. Hiramatsu and F. E. Osterloh, *Chem. Mater.*, 2004, **16**, 2509–2511.

18. J. Turkevich, P. C. Setevenson and J. Hillier, *Discuss. Faraday Soc.*, 1951, **11**, 55–75.
19. G. Frens, *Nature Phys. Sci.*, 1973, **241**, 20–22.
20. K. C. Grabar, R. G. Freeman, M. B. Hommer and M. J. Natan, *Anal. Chem.*, 1995, **67**, 735–743.
21. G. Schmid and U. Simon, *Chem. Commun.*, 2005, 697–710.
22. J. Petroski, M. H. Chou and C. Creutz, *Inorg. Chem.*, 2004, **43**, 1597–1599.
23. A. G. Kanaras, F. S. Kamounah, K. Schaumburg, C. J. Kiely and M. Brust, *Chem. Commun.*, 2002, 2294–2295.
24. M. Zheng, Z. Li and X. Huang, *Langmuir*, 2004, **20**, 4226–4235.
25. S. K. Bhargava, J. M. Booth, S. Agrawal, P. Coloe and G. Kar, *Langmuir*, 2005, **21**, 5949–5956.
26. D. V. Leff, L. Brandt and J. R. Heath, *Langmuir*, 1996, **12**, 4723–4730.
27. J. D. S. Newman and G. J. Blanchard, *Langmuir*, 2006, **22**, 5882–5887.
28. C. J. Zhong, W. X. Zhang, F. L. Leibowitz and H. H. Eichelberger, *Chem. Commun.*, 1999, 1211–1212.
29. M. M. Maye, W.-X. Zheng, F. L. Leibowitz, N. K. Ly and C. J. Zhong, *Langmuir*, 2000, **16**, 490–497.
30. J. B. Carroll, B. L. Frankamp, S. Srivastava and V. M. Rotello, *J. Mater. Chem.*, 2004, **14**, 690–694.
31. B. L. V. Prasad, S. I. Stoeva, C. M. Sorensen and K. J. Klabunde, *Langmuir*, 2002, **18**, 7515–7520.
32. B. L. V. Prasad, S. I. Stoeva, C. M. Sorensen and K. J. Klabunde, *Chem. Mater.*, 2003, **15**, 935–942.
33. T. K. Sau, A. Pal, N. R. Jana, Z. L. Wang and T. Pal, *J. Nanoparticle Res.*, 2001, **3**, 257–261.
34. H. Kurita, A. Takami and S. Koda, *Appl. Phys. Lett.*, 1998, **72**, 789–791.
35. A. C. Templeton, W. P. Wuelfing and R. W. Murray, *Acc. Chem. Res.*, 2000, **33**, 27–36.
36. R. Hong, J. M. Fernandez, H. Nakade, R. R. Arvizo, T. Emrick and V. M. Rotello, *Chem. Commun.*, 2006, 2347–2349.
37. A. K. Boal and V. M. Rotello, *J. Am. Chem. Soc.*, 2000, **122**, 734–735.
38. M. Zachary and V. Chechik, *Angew. Chem. Int. Ed.*, 2007, **46**, 3304–3307.
39. R. Levy, N. T. K. Thanh, R. C. Doty, I. Hussain, R. J. Nichols, D. J. Schiffrin, M. Brust and D. G. Fernig, *J. Am. Chem. Soc.*, 2004, **126**, 10076–10084.
40. G. H. Woehrle, L. O. Brown and J. E. Hutchison, *J. Am. Chem. Soc.*, 2005, **127**, 2172–2183.
41. S. Rucareanu, V. J. Gandubert and R. B. Lennox, *Chem. Mater.*, 2006, **18**, 4674–4680.
42. A. G. Tkachenko, H. Xie, D. Coleman, W. Golmm, J. Ryan, M. F. Anderson, S. Franzen and D. L. Feldheim, *J. Am. Chem. Soc.*, 2003, **125**, 4700–4701.
43. I. H. El-Sayed, X. Huang and M. A. El-Sayed, *Cancer Lett.*, 2006, **239**, 129–135.

44. S.-Y. Lin, Y.-T. Tsai, C.-C. Chen, C.-M. Lin and C.-H. Chen, *J. Phys. Chem. B*, 2004, **108**, 2134–2139.
45. K. Aslan and V. H. Perez-Luna, *Langmuir*, 2002, **18**, 6059–6065.
46. P. K. Jain, K. S. Lee, I. H. El-Sayed and M. A. El-Sayed, *J. Phys. Chem. B*, 2006, **110**, 7238–7248.
47. G. Mie, *Ann. Phys.*, 1908, **25**, 377–445.
48. K. L. Kelly, E. Coronado, L. L. Zhao and G. C. Schatz, *J. Phys. Chem. B*, 2003, **107**, 668–677.
49. S. Eustis and M. A. El-Sayed, *Chem. Soc. Rev.*, 2006, **35**, 209–217.
50. K.-H. Su, Q.-H. Wei, X. Zhang, J. J. Mock, D. R. Smith and S. Schultz, *Nano Lett.*, 2003, **3**, 1087–1090.
51. S. Link, Z. L. Wang and M. A. El-Sayed, *J. Phys. Chem. B*, 1999, **103**, 3529–3533.
52. X. Liu, M. Atwater, J. Wang and Q. Huo, *Colloid Surf. B: Biointerfaces*, 2006, **58**, 3–7.
53. P. K. Jain, I. H. El-Sayed and M. A. El-Sayed, *Nano Today*, 2007, **2**, 18–29.
54. J. Zheng, C. Zhang and R. M. Dickson, *Phys. Rev. Lett.*, 2004, **93**, 077402.
55. M. A. van Dijk, M. Lippitz and M. Orrit, *Acc. Chem. Res.*, 2005, **38**, 594–601.
56. J. R. Lakowicz, *Anal. Biochem.*, 2005, **337**, 171–194.
57. K. E. Sapsford, L. Berti and I. L. Medintz, *Angew. Chem. Int. Ed.*, 2006, **45**, 4562–4589.
58. E. Dulkeith, A. C. Morteani, T. Niedereichholz, T. A. Klar, J. Feldmann, S. A. Levi, F. C. J. M. van Veggel, D. N. Reinhoudt, M. Möller and D. I. Gittins, *Phys. Rev. Lett.*, 2002, **89**, 203002.
59. P. V. Kamat, S. Barazzouk and S. Hotchandani, *Angew. Chem. Int. Ed.*, 2002, **41**, 2764–2767.
60. B. M. Quinn, P. Liljeroth, V. Ruiz, T. Laaksonen and K. Kontturi, *J. Am. Chem. Soc.*, 2003, **125**, 6644–6645.
61. S. Antonello, A. H. Holm, E. Instuli and F. Maran, *J. Am. Chem. Soc.*, 2007, **129**, 9836–9837.
62. S. Srivastava, B. L. Frankamp and V. M. Rotello, *Chem. Mater.*, 2005, **17**, 487–490.
63. S.-Y. Lin, S.-W. Liu, C.-M. Lin and C.-H. Chen, *Anal. Chem.*, 2002, **74**, 330–335.
64. S.-Y. Lin, C.-H. Chen, M.-C. Lin and H.-F. Hsu, *Anal. Chem.*, 2005, **77**, 4821–4828.
65. S. O. Obare, R. E. Hollowell and C. J. Murphy, *Langmuir*, 2002, **18**, 10407–10410.
66. A. J. Reynolds, A. H. Haines and D. A. Russell, *Langmuir*, 2006, **22**, 1156–1163.
67. Y. J. Kim, R. C. Johnson and J. T. Hupp, *Nano Lett.*, 2001, **1**, 165–167.
68. C. C. Huang and H. T. Chang, *Chem. Commun.*, 2007, 1215–1217.

69. S.-Y. Lin, S.-H. Wu and C.-H. Chen, *Angew. Chem. Int. Ed.*, 2006, **45**, 4948–4951.

70. W. R. Yang, J. J. Gooding, Z. C. He, Q. Li and G. N. Chen, *J. Nanosci. Nanotechnol.*, 2007, **7**, 712–716.

71. S. Si, A. Kotal and T. K. Mandal, *J. Phys. Chem. C*, 2007, **111**, 1248–1255.

72. J.-S. Lee, M. S. Han and C. A. Mirkin, *Angew. Chem. Int. Ed.*, 2007, **46**, 4093–4096.

73. J. Liu and Y. Lu, *J. Am. Chem. Soc.*, 2003, **125**, 6642–6643.

74. J. W. Liu and Y. Lu, *Chem. Mater.*, 2004, **16**, 3231–3238.

75. J. W. Liu and Y. Lu, *J. Am. Chem. Soc.*, 2004, **126**, 12298–12305.

76. C. A. Mirkin, R. L. Letsinger, R. C. Mucic and J. J. Storhoff, *Nature*, 1996, **382**, 607–609.

77. R. Elghanian, J. J. Storhoff, R. C. Mucic, R. L. Letsinger and C. A. Mirkin, *Science*, 1997, **277**, 1078–1081.

78. J. Liu and Y. Lu, *J. Am. Chem. Soc.*, 2005, **127**, 12677–12683.

79. J. W. Liu and Y. Lu, *Anal. Chem.*, 2004, **76**, 1627–1632.

80. Y. Kubo, *Tetrahedron Lett.*, 2005, **46**, 4369–4372.

81. S. Watanabe, H. Seguchi, K. Yoshida, K. Kifune, T. Tadaki and H. Shiozaki, *Tetrahedron Lett.*, 2005, **46**, 8827–8829.

82. K. Aslan, J. R. Lakowicz and C. D. Geddes, *Anal. Biochem.*, 2004, **330**, 145–155.

83. K. Aslan, J. R. Lakowicz and C. D. Geddes, *Anal. Chem.*, 2005, **77**, 2007–2014.

84. J. Matsui, K. Akamatsu, S. Nishiguchi, D. Miyoshi, H. Nawafune, K. Tamaki and N. Sugimoto, *Anal. Chem.*, 2004, **76**, 1310–1315.

85. D. H. J. Bunka and P. G. Stockley, *Nat. Rev. Microbiol.*, 2006, **4**, 588–596.

86. J. Liu and Y. Lu, *Angew. Chem. Int. Ed.*, 2006, **45**, 90–94.

87. J. Liu and Y. Lu, *Adv. Mater.*, 2006, **18**, 1667–1671.

88. J. Liu, D. Mazumdar and Y. Lu, *Angew. Chem. Int. Ed.*, 2006, **45**, 7955–7959.

89. J. J. Storhoff, R. Elghanian, R. C. Mucic, C. A. Mirkin and R. L. Letsinger, *J. Am. Chem. Soc.*, 1998, **120**, 1959–1964.

90. R. A. Reynolds, C. A. Mirkin and R. L. Letsinger, *J. Am. Chem. Soc.*, 2000, **122**, 3795–3796.

91. Y. C. Cao, R. C. Jin, S. Thaxton and C. A. Mirkin, *Talanta*, 2005, **67**, 449–455.

92. C. S. Thaxton, D. G. Georganopoulou and C. A. Mirkin, *Clin. Chim. Acta*, 2006, **363**, 120–126.

93. H. X. Li and L. J. Rothberg, *J. Am. Chem. Soc.*, 2004, **126**, 10958–10961.

94. R. Chakrabarti and A. M. Klibanov, *J. Am. Chem. Soc.*, 2003, **125**, 12531–12540.

95. J. Li, X. Chu, Y. Liu, J.-H. Jiang, Z. He, Z. Zhang, G. Shen and R. Q. Yu, *Nucleic Acids Res.*, 2005, **33**, e168.

96. J. J. Storhoff, A. D. Lucas, V. Garimella, Y. P. Bao and U. R. Muller, *Nat. Biotechnol.*, 2004, **22**, 883–887.
97. U. Feldkamp and C. M. Niemeyer, *Angew. Chem. Int. Ed.*, 2006, **45**, 1856–1876.
98. H. Li and L. Rothberg, *Proc. Natl. Acad. Sci. USA*, 2004, **101**, 14036–14039.
99. M. S. Han, A. K. R. Lytton-Jean, B. K. Oh, J. Heo and C. A. Mirkin, *Angew. Chem. Int. Ed.*, 2006, **45**, 1807–1810.
100. M. S. Han, A. K. R. Lytton-Jean and C. A. Mirkin, *J. Am. Chem. Soc.*, 2006, **128**, 4954–4955.
101. H. Otsuka, Y. Akiyama, Y. Nagasaki and K. Kataoka, *J. Am. Chem. Soc.*, 2001, **123**, 8226–8230.
102. C. S. Tsai, T. B. Yu and C. T. Chen, *Chem. Commun.*, 2005, 4273–4275.
103. C. L. Schofield, R. A. Field and D. A. Russell, *Anal. Chem.*, 2007, **79**, 1356–1361.
104. N. Nath and A. Chilkoti, *Anal. Chem.*, 2002, **74**, 504–509.
105. N. Nath and A. Chilkoti, *Anal. Chem.*, 2004, **76**, 5370–5378.
106. C.-C. Huang, Y.-F. Huang, Z. Gao, W. Tan and H.-T. Chang, *Anal. Chem.*, 2005, **77**, 5735–5741.
107. H. Wei, B.-L. Li, J. Li, E.-K. Wang and S.-J. Dong, *Chem. Commun.*, 2007, DOI:10.1039/b707642h.
108. V. Pavlov, Y. Xiao, B. Shlyahovsky and I. Willner, *J. Am. Chem. Soc.*, 2004, **126**, 11768–11769.
109. Y. Xiao, V. Pavlov, S. Levine, T. Niazov, G. Markovitch and I. Willner, *Angew. Chem. Int. Ed.*, 2004, **43**, 4519–4522.
110. M. Brust, D. Bethell, D. J. Schiffrin and C. J. Kiely, *Adv. Mater.*, 1995, **7**, 795–799.
111. C. C. You, R. R. Arvizo and V. M. Rotello, *Chem. Commun.*, 2006, 2905–2907.
112. C. Guarise, L. Pasquato, V. De Filippis and P. Scrimin, *Proc. Natl. Acad. Sci. USA*, 2006, **103**, 3978–3982.
113. A. Laromaine, L. Koh, M. Murugesan, R. V. Ulijin and M. M. Stevens, *J. Am. Chem. Soc.*, 2007, **129**, 4156–4157.
114. X.-Y. Xu, M.-S. Han and C. A. Mirkin, *Angew. Chem. Int. Ed.*, 2007, **46**, 3468–3470.
115. Z. Wang, R. Levy, D. G. Fernig and M. Brust, *J. Am. Chem. Soc.*, 2006, **128**, 2214–2215.
116. Y. Choi, N.-H. Ho and C.-H. Tung, *Angew. Chem. Int. Ed.*, 2007, **46**, 707–709.
117. W. Zhao, W. Chiuman, J. C. F. Lam, M. A. Brook and Y. Li, *Chem. Commun.*, 2007, 3729–3731.
118. S. Chah, M. R. Hammond and R. N. Zare, *Chemistry & Biology*, 2005, **12**, 323–328.
119. T. Huang and R. W. Murray, *Langmuir*, 2002, **18**, 7077–7081.
120. C.-C. Huang and H.-T. Chang, *Anal. Chem.*, 2006, **78**, 8332–8338.

121. S.-J. Chen and H.-T. Chang, *Anal. Chem.*, 2004, **76**, 3727–3734.
122. X. R. He, H. B. Liu, Y. L. Li, S. Wang, Y. J. Li, N. Wang, J. C. Xiao, X. H. Xu and D. B. Zhu, *Adv. Mater.*, 2005, **17**, 2811.
123. B. I. Ipe, K. Yoosaf and K. G. Thomas, *J. Am. Chem. Soc.*, 2006, **128**, 1907–1913.
124. B. Dubertret, M. Calame and A. J. Libchaber, *Nat. Biotechnol.*, 2001, **19**, 365–370.
125. D. J. Maxwell, J. R. Taylor and S. M. Nie, *J. Am. Chem. Soc.*, 2002, **124**, 9606–9612.
126. P. C. Ray, A. Fortner and G. K. Darbha, *J. Phys. Chem. B*, 2006, **110**, 20745–20748.
127. L. Dyadyusha, H. Yin, S. Jaiswal, T. Brown, J. J. Baumberg, F. P. Booy and T. Melvin, *Chem. Commun.*, 2005, 3201–3203.
128. E. Oh, M.-Y. Hong, D. Lee, S.-H. Nam, H. C. Yoon and H.-S. Kim, *J. Am. Chem. Soc.*, 2005, **127**, 3270–3271.
129. E. Oh, D. Lee, Y. P. Kim, S. Y. Cha, D. B. Oh, H. A. Kang, J. Kim and H. S. Kim, *Angew. Chem. Int. Ed.*, 2006, **45**, 7959–7963.
130. C. C. You, O. R. Miranda, B. Gider, P. S. Ghosh, I. B. Kim, B. Erdogan, S. A. Krovi, U. H. F. Bunz and V. M. Rotello, *Nat. Nanotechnol.*, 2007, **2**, 318–323.
131. C. C. You, M. De, G. Han and V. M. Rotello, *J. Am. Chem. Soc.*, 2005, **127**, 12873–12881.
132. M. De, C. C. You, S. Srivastava and V. M. Rotello, *J. Am. Chem. Soc.*, 2007, **129**, 10747–10753.
133. R. L. Phillips, O. R. Miranda, C. C. You, V. M. Rotello and U. H. F. Bunz, *Angew. Chem. Int. Ed.* 2007, DOI: 10.1002/anie.200703369.
134. S. W. Thomas, G. D. Joly and T. M. Swager, *Chem. Rev.*, 2007, **107**, 1339–1386.
135. S. M. Nie and S. R. Emory, *Science*, 1997, **275**, 1102–1106.
136. R. F. Aroca, R. A. Alvarez-Puebla, N. Pieczonka, S. Sanchez-Cortez and J. V. Garcia-Ramos, *Adv. Colloid Interface Sci.*, 2005, **116**, 45–61.
137. F. Toderas, M. Baia, L. Baia and S. Astilean, *Nanotechnology*, 2007, **18**, 255702.
138. Y. C. Cao, R. Jin and C. A. Mirkin, *Science*, 2002, **297**, 1536–1540.
139. L. Sun, C. Yu and J. Irudayaraj, *Anal. Chem.*, 2007, **79**, 3981–3988.
140. Y. C. Cao, R. C. Jin, J. M. Nam, C. S. Thaxton and C. A. Mirkin, *J. Am. Chem. Soc.*, 2003, **125**, 14676–14677.
141. D. S. Grubisha, R. J. Lipert, H. Y. Park, J. Driskell and M. D. Porter, *Anal. Chem.*, 2003, **75**, 5936–5943.
142. E. Katz, I. Willner and J. Wang, *Electroanalysis*, 2004, **16**, 19–44.
143. A. M. Yu, Z. J. Liang, J. H. Cho and F. Caruso, *Nano Lett.*, 2003, **3**, 1203–1207.
144. C. R. Raj, T. Okajima and T. Ohsaka, *J. Electroanal. Chem.*, 2003, **543**, 127–133.

145. F. P. Zamborini, M. C. Leopold, J. F. Hicks, P. J. Kulesza, M. A. Malik and R. W. Murray, *J. Am. Chem. Soc.*, 2002, **124**, 8958–8964.
146. H. Wohltjen and A. W. Snow, *Anal. Chem.*, 1998, **70**, 2856–2859.
147. S. D. Evans, S. R. Johnson, Y. L. Cheng and T. Shen, *J. Mater. Chem.*, 2000, **10**, 183–188.
148. Y. Joseph, I. Besnard, M. Rosenberger, B. Guse, H.-G. Nothofer, J. M. Wessels, U. Wild, A. Knop-Gericke, D.-S. Su, R. Schlögl, A. Yasuda and T. Vossmeyer, *J. Phys. Chem. B*, 2003, **107**, 7406–7413.
149. L. Y. Wang, N. N. Kariuki, M. Schadt, D. Mott, J. Luo, C. J. Zhong, X. J. Shi, C. Zhang, W. B. Hao, S. Lu, N. Kim and J. Q. Wang, *Sensors*, 2006, **6**, 667–679.
150. L. Y. Wang, X. J. Shi, N. N. Kariuki, M. Schadt, G. R. Wang, Q. Rendeng, J. Choi, J. Luo, S. Lu and C. J. Zhong, *J. Am. Chem. Soc.*, 2007, **129**, 2161–2170.
151. N. Krasteva, I. Besnard, B. Guse, R. E. Bauer, K. Müllen, A. Yasuda and T. Vossmeyer, *Nano Lett.*, 2002, **2**, 551–555.
152. A. Singh, S. Hede and M. Sastry, *Small*, 2007, **3**, 466–473.
153. A. N. Shipway, M. Lahav, R. Blonder and I. Willner, *Chem. Mater.*, 1999, **11**, 13–15.
154. M. Lahav, A. N. Shipway and I. Willner, *J. Chem. Soc., Perkin Trans. 2*, 1999, 1925–1931.
155. M. Lahav, A. N. Shipway, I. Willner, M. B. Nielsen and J. F. Stoddart, *J. Electroanal. Chem.*, 2000, **482**, 217–221.
156. M. Lahav, R. Gabai, A. N. Shipway and I. Willner, *Chem. Commun.*, 1999, 1937–1938.
157. R. Blonder, L. Sheeney and I. Willner, *Chem. Commun.*, 1998, 1393–1394.
158. A. B. Kharitonov, A. N. Shipway and I. Willner, *Anal. Chem.*, 1999, **71**, 5441–5443.
159. J. F. Geng, M. D. R. Thomas, D. S. Shephard and B. F. G. Johnson, *Chem. Commun.*, 2005, 1895–1897.
160. D. T. Thompson, *Nanotoday*, 2007, **2**(4), 40–43.
161. C. R. Raj and B. K. Jena, *Chem. Commun.*, 2005, 2005–2007.
162. B. K. Jena and C. R. Raj, *Chem. Eur. J.*, 2006, **12**, 2702–2708.
163. B. K. Jena and C. R. Raj, *Langmuir*, 2007, **23**, 4064–4070.
164. A. L. Crumbliss, S. C. Perine, J. Stonehürner, K. R. Tubergen, J. G. Zhao and R. W. Henkens, *Biotech. Bioeng.*, 1992, **40**, 483–490.
165. X. X. Xu, S. Q. Liu, B. Li and H. X. Ju, *Anal. Lett.*, 2003, **36**, 2427–2442.
166. O. Shulga and J. R. Kirchhoff, *Electrochem. Commun.*, 2007, **9**, 935–940.
167. S. Bharathi and M. Nogami, *Analyst*, 2001, **126**, 1919–1922.
168. C. C. You, S. S. Agasti, M. De, M. J. Knapp and V. M. Rotello, *J. Am. Chem. Soc.*, 2006, **128**, 14612–14618.
169. P. Pandey, S. P. Singh, S. K. Arya, V. Gupta, M. Datta, S. Singh and B. D. Malhotra, *Langmuir*, 2007, **23**, 3333–3337.
170. Y. Xiao, F. Patolsky, E. Katz, J. F. Hainfeld and I. Willner, *Science*, 2003, **299**, 1877–1881.

171. M. T. Castaneda, S. Alegret and A. Merkoci, *Electroanalysis*, 2007, **19**, 743–753.
172. S. J. Park, T. A. Taton and C. A. Mirkin, *Science*, 2002, **295**, 1503–1506.
173. M. Ozsoz, A. Erdem, K. Kerman, D. Ozkan, B. Tugrul, N. Topcuoglu, H. Ekren and M. Taylan, *Anal. Chem.*, 2003, **75**, 2181–2187.
174. H. Cai, Y. Q. Wang, P. G. He and Y. H. Fang, *Anal. Chim. Acta*, 2002, **469**, 165–172.
175. J. Wang, J. H. Li, A. J. Baca, J. B. Hu, F. M. Zhou, W. Yan and D. W. Pang, *Anal. Chem.*, 2003, **75**, 3941–3945.
176. A. J. Baca, F. M. Zhou, J. Wang, J. B. Hu, J. H. Li, J. X. Wang and Z. S. Chikneyan, *Electroanalysis*, 2004, **16**, 73–80.
177. J. Zhang, S. P. Song, L. Y. Zhang, L. H. Wang, H. P. Wu, D. Pan and C. Fan, *J. Am. Chem. Soc.*, 2006, **128**, 8575–8580.
178. D. Li, Y. Yan, A. Wieckowska and I. Willner, *Chem. Commun.*, 2007, 3544–3546.
179. O. D. Velev and E. W. Kaler, *Langmuir*, 1999, **15**, 3693–3698.
180. M. Dequaire, C. Degrand and B. Limoges, *Anal. Chem.*, 2000, **72**, 5521–5528.
181. X. Mao, J. Jiang, J. Chen, Y. Huang, G. Shen and R. Yu, *Anal. Chim. Acta*, 2006, **557**, 159–163.
182. J. Homola, S. S. Yee and G. Gauglitz, *Sens. Actuators B*, 1999, **54**, 3–15.
183. J. Wang, *Small*, 2005, **1**, 1036–1043.
184. J. Matsui, K. Akamatsu, N. Hara, D. Miyoshi, H. Nawafune, K. Tamaki and N. Sugimoto, *Anal. Chem.*, 2005, **77**, 4282–4285.
185. L. He, M. D. Musick, S. R. Nicewarner, F. G. Salinas, S. J. Benkovic, M. J. Natan and C. D. Keating, *J. Am. Chem. Soc.*, 2000, **122**, 9071–9077.
186. S. P. Fang, H. J. Lee, A. W. Wark and R. M. Corn, *J. Am. Chem. Soc.*, 2006, **128**, 14044–14046.
187. X. Yao, X. Li, F. Toledo, C. Zurita-Lopez, M. Gutova, J. Momand and F. M. Zhou, *Anal. Biochem.*, 2006, **354**, 220–228.
188. R. C. Bailey, J. M. Nam, C. A. Mirkin and J. T. Hupp, *J. Am. Chem. Soc.*, 2003, **125**, 13541–13547.
189. L. R. Hirsch, J. B. Jackson, A. Lee, N. J. Halas and J. L. West, *Anal. Chem.*, 2003, **75**, 2377–2381.
190. L. A. Lyon, M. D. Musick and M. J. Natan, *Anal. Chem.*, 1998, **70**, 5177–5183.
191. U. Pieper-Furst, W. F. M. Stocklein and A. Warsinke, *Anal. Chim. Acta*, 2005, **550**, 69–76.
192. K. A. Marx, *Biomacromolecules*, 2003, **4**, 1099–1120.
193. L. Han, D. R. Daniel, M. M. Maye and C. J. Zhong, *Anal. Chem.*, 2001, **73**, 4441–4449.
194. J. W. Grate, D. A. Nelson and R. Skaggs, *Anal. Chem.*, 2003, **75**, 1868–1879.
195. N. Krasteva, Y. Fogel, R. E. Bauer, K. Mullen, N. Matsuzawa, A. Yasuda and T. Vossmeyer, *Adv. Funct. Mater.*, 2007, **17**, 881–888.

196. X. L. Mao, L. J. Yang, X. L. Su and Y. B. Li, *Biosensors & Bioelectronics*, 2006, **21**, 1178–1185.
197. L. Lin, H. Q. Zhao, J. R. Li, J. A. Tang, M. X. Duan and L. Jiang, *Biochem. Biophys. Res. Commun.*, 2000, **274**, 817–820.
198. X.-C. Zhou, S. J. O'shea and S. F. Y. Li, *Chem. Commun.*, 2000, 953–954.
199. F. Patolsky, K. T. Ranjit, A. Lichtenstein and I. Willner, *Chem. Commun.*, 2000, 1025–1026.
200. I. Willner, F. Patolsky, Y. Weizmann and B. Willner, *Talanta*, 2002, **56**, 847–856.
201. L. B. Nie, Y. Yang, S. Li and N. Y. He, *Nanotechnology*, 2007, **18**, 305501.
202. T. Liu, J. Tang, M. M. Han and L. Jiang, *Biochem. Biophys. Res. Commun.*, 2003, **304**, 98–100.
203. Y. Weizmann, F. Patolsky and I. Willner, *Analyst*, 2001, **126**, 1502–1504.
204. T. Liu, J. Tang and L. Jiang, *Biochem. Biophys. Res. Commun.*, 2002, **295**, 14–16.
205. M. Su, S. U. Li and V. P. Dravid, *Appl. Phys. Lett.*, 2003, **82**, 3562–3564.
206. N. H. Kim, T. J. Baek, H. G. Park and G. H. Seong, *Anal. Sci.*, 2007, **23**, 177–181.
207. J. M. Nam, C. S. Thaxton and C. A. Mirkin, *Science*, 2003, **301**, 1884–1886.
208. D. G. Georganopoulou, L. Chang, J. M. Nam, C. S. Thaxton, E. J. Mufson, W. L. Klein and C. A. Mirkin, *Proc. Natl. Acad. Sci. USA*, 2005, **102**, 2273–2276.
209. S. I. Stoeva, J. S. Lee, J. E. Smith, S. T. Rosen and C. A. Mirkin, *J. Am. Chem. Soc.*, 2006, **128**, 8378–8379.
210. B. K. Oh, J. M. Nam, S. W. Lee and C. A. Mirkin, *Small*, 2006, **2**, 103–108.
211. J. M. Nam, A. R. Wise and J. T. Groves, *Anal. Chem.*, 2005, **77**, 6985–6988.
212. J. M. Nam, K.-J. Jang and J. T. Groves, *Nat. Protocols*, 2007, **2**, 1438–1444.
213. J. M. Nam, S. I. Stoeva and C. A. Mirkin, *J. Am. Chem. Soc.*, 2004, **126**, 5932–5933.
214. S. I. Stoeva, J. S. Lee, C. S. Thaxton and C. A. Mirkin, *Angew. Chem. Int. Ed.*, 2006, **45**, 3303–3306.

CHAPTER 3

Resistive-pulse Sensing and On-chip Artificial Pores for Biological Sensing

OMAR A. SALEH[a] AND LYDIA L. SOHN[b]

[a] Materials Department and Biomolecular Science and Engineering Program, University of California, Santa Barbara, Santa Barbara, CA 93106-5050, USA; [b] Department of Mechanical Engineering, University of California, Berkeley, Berkeley, CA 94720-1740, USA

3.1 Introduction

Resistive-pulse sensing (RPS) is a powerful technique that can be used to analyze the size and concentration of particles dispersed in a solution. RPS consists of measuring the electrical resistance of an aperture (or pore) that connects two fluid-filled reservoirs (Figure 3.1). Particles in solutions pass from one reservoir to another through the aperture due to either an applied force (typically an electric field or a hydrodynamic drag) or simply by diffusion. When a particle inhabits the aperture, it displaces conducting fluid, and thus causes a transient increase, or pulse, in the aperture's electrical resistance. The magnitude of this resistance change is indicative of the size of the particle, and the frequency at which pulses occur is related to the particle concentration. The method of RPS was invented in 1953 by W. H. Coulter.[1] His "Coulter counter" contained an aperture typically tens of microns in diameter; it is still widely used to measure the size distributions of biological cells.[2,3]

Successful RPS requires the size of the analyzed particle to be of the same order as the size of the pore. Because of this constraint, inventing new methods

Nano and Microsensors for Chemical and Biological Terrorism Surveillance
Edited by Jeffrey B.-H. Tok
© Royal Society of Chemistry, 2008
Published by the Royal Society of Chemistry, www.rsc.org

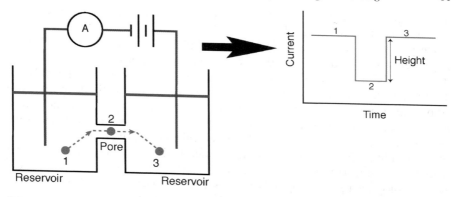

Figure 3.1 Schematic of the basic components of a resistive-pulse sensor. A particle passing through the pore causes a pulse in the measured electrical current; the height of the pulse is proportional to the size of the particle.

to make smaller pores (and thus measure smaller particles) has been a major trend in the field. The first step in this direction was taken in the 1970s, when R. W. Deblois and C. P. Bean invented the so-called "Nanopar" device,[4] which consisted of a plastic sheet separating the reservoirs and contained a single pore 0.5–3 µm in diameter and several microns in length. The pores were realized using the "Nuclepore" process, which involved irradiating the plastic sheet with high-energy nuclear particles and then using a basic solution to etch pores along the damage tracks left by the particles. Deblois and Bean further recognized that the homogeneous electric field in a long, thin pore (as compared to the field in the relatively short and wide pore used by Coulter) allows greater analytical precision in relating resistive pulse height to particle size. They used their device to 1) measure precisely colloidal particles with diameters ≥ 100 nm,[4] 2) analyze sperm cells,[5] and 3) characterize viral particles as small as 60 nm in diameter based on differences of both size and charge.[6–8] Later, von Schulthess and Benedek collaborated with Deblois to use the Nanopar device to investigate the agglutination of antigen-coated particles by antibody.[9,10]

In 1994, the applicable length scale of RPS was drastically decreased when Bezrukov *et al.* successfully utilized an ion channel as a molecular Coulter counter.[11,12] In their device, the reservoirs were separated by a lipid bilayer into which an ion channel, extracted from a cell membrane, inserted itself. Measurements of the current through the nanometer-scale opening in the ion channel revealed the passage of single molecules contained in solution. Perhaps the most important facet of this work was the introduction of analyte specificity to RPS. Previously, RPS only revealed geometric attributes, such as size and shape, of the particles passing through the aperture. Ion channels, on the other hand, have functional chemical groups on the interiors, thus providing opportunities for chemical interaction between the particles and the pore. Such interactions typically are revealed by changes in the width of the resistive pulses. Currently, many groups are exploiting the capabilities of ion channels in an attempt to use them as biosensors that can be engineered to have sensitivities to many different molecules.[12–18]

Despite the success of several groups in utilizing ion channels as resistive-pulse sensors of small molecules, the creation of a robust ion-channel-based device has been hampered by problems with the stability of the lipid bilayer. These concerns have led to several groups to work on new methods to fabricate stable "artificial" pores (compared to the "natural" protein ion channels) that are based in solid materials. Compared to the natural pores, the artificial pores must share at least one of their attributes: chemical functionality or nanometer-scale openings. To create chemically functionalized devices, groups have worked on using techniques of surface chemistry to attach covalently inter-acting molecules to the inside of a metal-or silica-based pore.[19,20] Two strate-gies have recently been used to create nanometer-scale devices. Sun and Crooks used multi-walled carbon nanotubes as templates for pores in metallic films.[21] They successfully created ~150nm diameter devices and claimed that 10 nm diameter pores are attainable with their method. More impressively, while using an ion beam to open a ~100 nm diameter hole in a thin silicon nitride membrane, Li *et al.* discovered a re-deposition method that allowed the crea-tion of 1.8 nm diameter pores.[22] While apparently quite fragile, these pores were successfully used to detect DNA molecules of 500 base pairs length.[23]

In contrast to the two main types of pores utilized today – ion channels suspended in lipid bilayers,[11,16,18,24,25] and molecular-scaled holes in silicon nitride[23,26–30] – we have developed a fundamentally different artificial pore that is completely chip-based and fabricated with great ease and control using common micro- and nano-fabrication and soft-lithography[31] techniques. As will be described in the following sections, our rapid and simple fabrication process allows the creation of stable pores of diameters from 200 nm to 1 μm (and beyond). Further, our process permits a large amount of flexibility in design, and is thus ideally suited for future integration with other microfluidic technologies.

3.2 Device Fabrication

Utilizing modern micro- and nano-fabrication techniques allow us to fabricate our pore-based devices rapidly, reproducibly, and with significant flexibility in design. We use both optical and electron-beam lithographies to create patterns that are either transferred to a substrate through metal deposition or reactive ion-etching (RIE) or used as molds for polydimethylsiloxane (PDMS), a curable rubber. These transferred patterns form the pore, reservoirs, and electrodes of a finished device.

We have fabricated two basic types of devices: one in which the pore and reservoirs are etched into a quartz substrate, and one in which the pore and reservoirs are embedded into a slab of PDMS. The former is used in the experiments described in Section 3.3, and the latter in the experiments described in Sections 3.4 and 3.5. Here, we describe the basic steps in the construction of each type of device; specialized features of particular devices will be described in Sections 3.3–3.5.

3.2.1 Etched Quartz Devices

The device shown in Figure 3.2 is fabricated in multiple stages. Each stage consists of lithographic pattern generation, followed by pattern transfer onto a quartz substrate using RIE or metal deposition and liftoff. The first stage is the fabrication of the pore. A line is patterned onto the substrate using either photolithography (PL) for line widths $>1\,\mu$m, or electron-beam lithography (EBL) for line widths between 100 and 500 nm, and then etched into the quartz using a CHF_3 RIE. The substrate subsequently undergoes a second stage of PL and RIE to define two reservoirs that are 1) considerably larger than the pore (3.5 μm deep and 10 μm wide), 2) separated by 10 μm, and 3) connected to each other by the previously-defined channel (see Figure 3.2). The length of the pore is thus defined in this second stage by the separation between the two reservoirs.

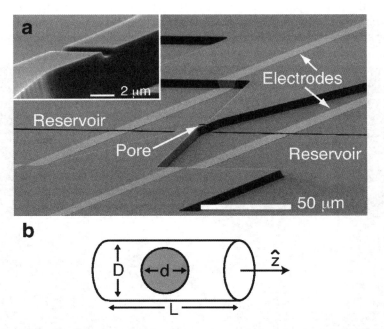

Figure 3.2 (a) Scanning electron micrograph of our on-chip artificial pore. The reservoirs (3.5 μm deep) and the inner electrodes, which control the voltage applied but pass no current, are only partially shown. The outer electrodes, which inject current into the solution, are not visible in this micrograph. Inset: magnified view of the pore that has dimensions $5.1 \times 1.5 \times 1.0\,\mu$m^3. (b) A schematic diagram of a spherical particle of diameter d in a pore of diameter D and length L. From ref. 34.

The final stage consists of patterning four electrodes across the reservoirs, followed by two depositions of 50/250 Å Ti/Pt in an electron-beam evaporator with the sample positioned ±45° from normal to the flux of metal to ensure that the electrodes are continuous down both walls of the reservoirs. At this point, the fabrication of the substrate is complete.

The device is sealed on top with a silicone-coated (Sylgard 184, Dow Corning Corp.) glass coverslip before each measurement. Prior to sealing, both the silicone and the substrate are oxidized in a dc plasma to ensure the hydrophilicity[32] of the reservoir and pore and to strengthen the seal[33] to the quartz substrate. After each measurement, the coverslip is removed and discarded, and the substrate is cleaned by chemical and ultrasonic methods. Thus, each device can be reused many times.

3.2.2 Molded PDMS Devices

Fabrication of a molded PDMS device begins with creating a master (Figure 3.3), from which a slab of PDMS will eventually be cast *via* standard micromolding techniques.[31] The master is created in two steps. First, a raised line that will become the pore is patterned using PL or EBL. Next, PL is used to pattern an SU-8 photoresist on the substrate to form the negatives of the reservoirs. Following standard micromolding techniques,[31] we pour PDMS over the master and

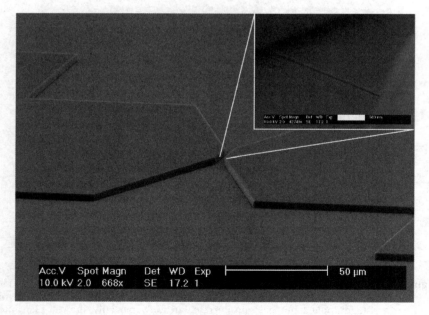

Figure 3.3 Scanning electron micrograph of a master containing an electron-beam lithography (EBL) defined pore. The image shows the raised resist structures that will be cast as reservoirs, while the inset shows an EBL-defined resist line that will be cast as the pore. The scale bar in the inset is 500 nm.

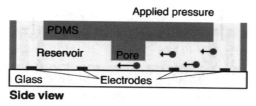

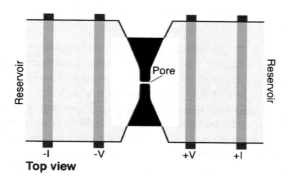

Figure 3.4 Schematic top and side views of a PDMS-based device. Incorporated into the top view is an optical image of an actual sealed device containing a 7-μm long, 1-μm wide pore. From ref. 42.

cure it at 80 °C for at least 24 hrs to ensure stable mechanical properties. The PDMS slab is then removed from the master and access holes are cored using a needle. Following thorough cleaning, the PDMS slab is permanently sealed, *via* a heat treatment, to a glass substrate that has previously defined Pt electrodes. Figure 3.4 shows a completed device.

3.2.3 Measurement

Once complete, the quartz or the PDMS-based device can be wet, *via* capillary action, with the solution to be studied. The suspended particles must be made to flow through the pore and the electrical current through the pore subsequently measured.

3.2.3.1 Inducing Particle Flow

There are two basic methods for inducing the particles in solution to pass through the pore: electrophoresis or pressure-induced fluid flow. In practice, electrophoresis is a "cleaner" method: any foreign objects in solution that might clog the pore will generally not have a very high charge, thus only the (charged) particles of interest will move. However, not all particles of interest are charged, and fluid flow must be used in those cases.

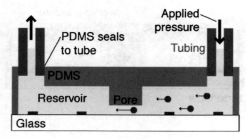

Figure 3.5   Method for pumping fluid through a PDMS-based device. From ref. 42.

Electrophoresis in the pore devices is readily achievable, as the electrodes are already fabricated on the devices for the purpose of measuring the current. The same voltage used to measure the current is used to manipulate the particles. Typically, an application of 0.5 V will cause DNA molecules to transit an EBL-defined pore 4 μm long in a few milliseconds, or 500 nm diameter carboxyl-coated colloids to transit a PL-defined pore 10 μm long in 50–100 ms. A by-product of the application of voltage in microfluidic channels is electro-osmosis. In our devices, the walls of the (quartz or PDMS) channels tend to carry a negative charge, so the fluid will carry a positive charge – this allows us to distinguish the electrophoretic motion of negatively charged particles from the oppositely-directed electro-osmotic flow of a positively-charged solution. While electro-osmotic effects have been seen in our devices when utilizing very high voltages (> 10 V), they do not appear to be significant at the voltages we used in a typical measurement.

Pressure-induced fluid flow is used only with PDMS-based devices. Once the device is wet with solution, pressure can be applied to either reservoir using a commercial microfluidic pump (Fluidigm, Inc.). The pump is connected to the device through Tygon tubing that is inserted into the access holes in the PDMS slab (Figure 3.5). 7–14 kPa (1–2 psi) is typically applied during measurements, inducing a 500-nm diameter colloid to transit the pore in a few hundred microseconds. Clogs in the pore occur occasionally due to either colloidal aggregates or unwanted objects, but they can usually be cleared by application of higher pressures (up to 2 psi), without any effect on the PDMS/coverslip seal.

3.2.3.2   Electronic Measurement

The sensitivity of our device relies upon the relative sizes of the pore and the particle to be measured. The resistance of a pore R_p increases by δR_p when a particle enters since the particle displaces conducting fluid. δR_p can be estimated[3] for a pore aligned along the z-axis (Figure 3.2) by

$$\delta R_p = \rho \int \frac{dz}{A(z)} - R_p \qquad (3.1)$$

where $A(z)$ represents the successive cross sections of the pore containing a particle, and ρ is the resistivity of the solution. For a spherical particle of diameter d in a pore of diameter D and length L, the relative change in resistance is

$$\frac{\delta R_p}{R_p} = \frac{D}{L} \left| \frac{\arcsin(d/D)}{(1-(d/D)^2)^{1/2}} - \frac{d}{D} \right| \tag{3.2}$$

Equations (3.1) and (3.2) assume that the current density is uniform across the pore, and thus is not applicable for cases where the cross section $A(z)$ varies quickly, *i.e.* when $d \ll D$. For that particular case, Deblois and Bean[4] formulated an equation for δR_p based on an approximate solution to the Laplace equation:

$$\frac{\delta R_p}{R_p} = \frac{d^3}{LD^2} \left| \frac{D^2}{2L^2} + \frac{1}{\sqrt{1+(D/L)^2}} \right| F\left(\frac{d^3}{D^3}\right) \tag{3.3}$$

where $F(d^3/D^3)$ is a numerical factor that accounts for the bulging of the electric field lines into the pore wall. When employing Equation (3.3) to predict resistance changes, we find an effective value for D, which takes into account the cross-sectional area of our square pore with that of a circular pore.

If R_p is the dominant resistance of the measurement circuit, then relative changes in the current I are equal in magnitude to the relative changes in the resistance, $|\delta I/I| = |\delta R_p/R_p|$, and Equations (3.2) and (3.3) can both be directly compared to measured current changes. This comparison is disallowed if R_p is similar in magnitude to other series resistances, such as the electrode/fluid interfacial resistance, $R_{e/f}$, or the resistance R_u of the reservoir fluid between the inner electrodes and the pore. We completely remove $R_{e/f}$ from the electrical circuit by performing a four-point measurement of the current (Figures 3.2 and 3.4). We minimize R_u by placing the inner electrodes close to the pore (50 μm away on either side), and by designing the reservoir with a cross section much larger than that of the pore. For a pore of dimensions 10.5 μm by 1.04 μm², we measured $R_p = 36$ MΩ, in good agreement with the 39 MΩ value predicted by the pore geometry and the solution resistivity. This confirms that we have removed R_u and $R_{e/f}$ from the circuit.

3.3   Quantitative Measurement of Polydisperse Colloidal Solutions

For proof-of-principle, we have measured solutions of negatively-charged (carboxyl-coated) latex colloids (Interfacial Dynamics, Inc.) whose diameters

ranged from 87 to 640 nm.[34] All colloids were suspended in a solution of 5×
concentration Tris-Borate-EDTA (TBE) buffer with a resistivity of 390 Ω cm
and pH 8.2. To reduce adhesion of the colloids to the reservoir and pore walls,
we added 0.05% volume to volume (v/v) of the surfactant Tween 20 to every
solution. The colloidal suspensions were diluted significantly from stock concentrations to avoid jamming of colloids in the pore; typical final concentrations were ~10^8 particles/mL.

Figure 3.6 shows typical measurements of the normalized current $\delta I/I$ *vs.*
time. Each downward pulse in the figure represents a single colloid transiting
the pore. Events in which two colloids simultaneously inhabit the pore are seen,
but are easily differentiated from single particle events by their anomalous pulse
heights and widths, as shown in Figure 3.7. As shown in Figure 3.6a, the
passage of 87-nm particles is easily detected in an 8.3-μm long pore of cross
section 0.16 μm². Measurements of a polydisperse solution of colloids are
shown in Figure 3.6b: there are clear differences in the pulse heights caused by
particles of different diameters.

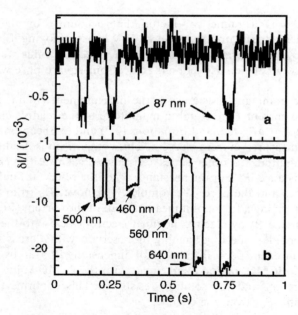

Figure 3.6 Relative changes in baseline current $\delta I/I$ *vs.* time for (a) a monodisperse
solution of 87 nm diameter latex colloids measured with an EBL-defined
pore of length 8.3 μm and cross section 0.16 μm², and (b) a polydisperse
solution of latex colloids with diameters 460, 500, 560 and 640 nm measured with a PL-defined pore of length 0.5 μm and cross section 1.2 μm².
Each downward pulse represents an individual particular entering the
pore. The four distinct pulse heights in (b) correspond as labeled to the
four different colloid diameters. From ref. 34.

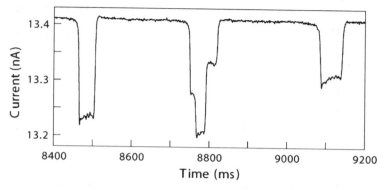

Figure 3.7 A current trace showing an event (center pulse) resulting from two colloids simultaneously inhabiting the pore; also shown are two single-colloid events (left and right pulses). Such two-colloid events are easily removed from the data analysis due to their anomalous shape.

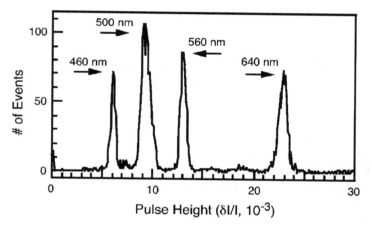

Figure 3.8 A histogram of pulse heights resulting from measuring the polydisperse solution shown in Figure 3.6b. The resolution of this particular device is ±10 nm in diameter for the particles measured. From ref. 34.

Figure 3.8 shows a histogram of ~3000 events measured for the same polydisperse solution shown in Figure 3.6b. The histogram emphasizes the ability of the device to separate colloids of different diameters. The widths of the peaks in the histogram represent the resolution of the device in sizing particles. We find that the response of the device for each type of colloid varies by 2–4%, which approaches the intrinsic variation of the diameter quoted by the colloid manufacturer.

We used a device whose pore size was 10.5 μm by 1.05 μm² to measure colloids ranging from 190 to 640 nm in diameter. Figure 3.9 shows a comparison

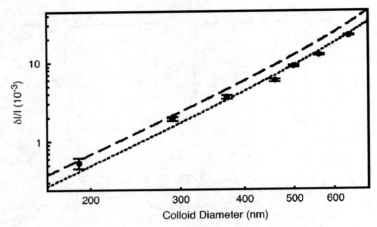

Figure 3.9 Comparison of measured $\delta I / I$ values (circles) with those predicted by Equation (3.2) (dotted line) and Equation (3.3) (dashed line). The measured data were taken over several runs on a single PL-defined pore of length 10.6 μm and cross section 1.04 μm². Error bars for the larger colloid sizes are obscured by the size of the plotted point. As the colloid diameter increases, there is a transition from agreement with Equation (3.3) to Equation (3.2). This reflects the fact that the derivation of Equation (3.3) assumes the colloid diameter d is much less than the pore diameter D; conversely Equation (3.2) relies on an assumption that holds only as d approaches D, and breaks down for smaller colloids. From ref. 34.

between the measured mean pulse heights and those predicted by Equations (3.2) and (3.3). As shown, there is excellent agreement between the measured and calculated values, with the measured error insignificant compared to the range of pulse heights. In addition, the measurements more closely follow Equation (3.3) for small d and Equation (3.2) for larger d, as was anticipated in the derivation of those equations.

3.4 Immunological Sensing

Antibodies can be powerful and flexible tools because of their natural ability to bind to virtually any molecule and because of the modern ability to produce specific types in large quantities. These traits have led to the development of a number of important immunosensing techniques in which antibodies of a desired specificity are used to test for the presence of a given antigen.[35–38] For example, radioimmunoassays (RIA) have been employed in clinical settings to screen for such viruses as hepatitis.[39] An integral part of all immunosensing technologies is the ability to detect the binding of antibody to antigen. To accomplish this, most common immunoassays require the labeling of the antibody using fluorescence, radioactivity, or enzyme activity. However, the

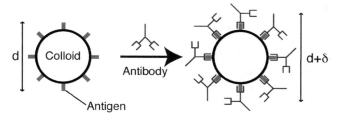

Figure 3.10   Schematic diagram of the binding of antibody to colloids. As shown, the strategy we employ to detect the binding of antibody to antigen-coated colloids involves measuring the colloid diameter increase from d to $d + \delta$ due to the volume added by the bound antibody.

need to bind chemically a label to the antibody adds to the time and cost of developing and employing these technologies.

Here, we describe how we have used our PDMS-based on-chip resistive pulse sensor to perform immunoassays. We show that we can detect the increase in diameter of a latex colloid upon binding to an unlabeled specific antibody (Figure 3.10). We have employed this novel technique to perform two important types of immunoassays: an inhibition assay, in which we detect the presence of an antigen by its ability to disrupt the binding of antibody to the colloid; and a sandwich assay, in which we successively detect the binding of each antibody in a two-site configuration.

Previous particle-counting-based immunoassays have used optical scattering or resistive-pulse methods to detect the aggregates formed when the antibody crosslinks antigen-coated colloids.[9,10,40] However, relying on crosslinking as a general binding probe is limiting since it requires a free ligand with at least two binding sites. In contrast, our method is more general, since it relies only on the added volume of bound ligand and does not place any limitations on the ligand's functionality. While our device cannot as of yet perform the kinetic analyses of which surface plasmon resonance (SPR) techniques[41] are capable, it represents a rapid, inexpensive, and compact alternative to SPR for end-point analysis of biological reactions.

3.4.1   Experimental Details

For the immunoassays described here, PDMS-based devices are used.[42] These devices have pores 7–9 μm in length and are 0.9–1.2 μm in diameter. Two types of colloids are used in these experiments: an experimental colloid, on which the binding reactions of interest occur, and a reference colloid, which is used to calibrate the device as will be described below. The experimental particle is a streptavidin-coated latex colloid of mean diameter ~ 510 nm, while the reference particle is a sulfate-coated latex colloid of mean diameter 470 nm. The colloids are mixed in solutions of PBS at pH 7.3 containing the surfactants BSA (0.2 mg/mL) and pluronics F127 (0.05% by volume).

The colloids are driven through the pore using pressure-driven flow (see Section 3.2.3.1). When applying 1–2 psi, each colloid passes through the pore in a few hundred microseconds; this time is long enough to establish a stable square pulse. Figure 3.11 shows a typical current trace that records several colloids passing through the pore; the inset magnifies two pulses to reveal their square shape.

A single experimental run consists of at least several hundred of each type of colloid passing through the pore. We use the experimental results from 3–5 different devices to find the mean diameter of a population of colloids suspended in a particular solution. The dominant source of error in our measurements is the intrinsic distribution of the streptavidin colloids' diameters, with smaller contributions from the spread in diameter of the reference colloids and the electrical noise in the current measurement (which stems mainly from the Johnson noise across the pore).

Because PDMS is a flexible material that can be distorted upon permanent bonding, we must calibrate each pore. Since the colloids to be measured (~ 510 nm) are comparable to the size of the pore (~ 900 nm), *i.e.* $d \sim D$, the relative height of a pulse $\delta I/I$ is described by Equation (3.2) .[3,4,34,43,44] Using this equation, we can determine d for each streptavidin colloid measured if we know the dimensions of the pore. We can directly measure L under an optical

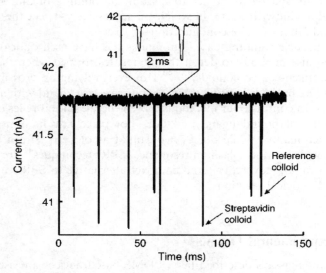

Figure 3.11 A typical measurement of the current across a pore as different colloids pass through it. Each downward pulse corresponds to a single colloid transiting the pore. There is a clear difference in pulse magnitude as a result of the difference in size of the streptavidin colloids as compared to the reference colloids. This difference allows us to separate the pulses for pore calibration (see text). The inset shows an expanded view of two pulses. As shown, they are well resolved in time and consequently allow an unambiguous measurement of the pulse height. The data shown were taken with an applied voltage of 0.4 V and a pressure of ~ 6.9 kPa. From ref. 42.

microscope. However, we cannot directly measure the pore's diameter D; instead, we perform a calibration by adding a reference colloid of known diameter (a 470 nm diameter sulfate-coated latex colloid) to each solution of streptavidin colloids. The absolute difference in diameter (470 nm to 510 nm) between the two types of colloids results in a clear difference in the pulse heights (Figure 3.11) and, consequently, we can determine easily which size colloid produced each pulse. We use the values of $\delta I/I$ arising from the reference colloids, along with the known values of L and d, to invert numerically Equation (3.2) to thus determine the pore diameter D. Once this is accomplished, we use Equation (3.2) once again to correlate the magnitude of each pulse to the diameter of the streptavidin colloid that produced it.

3.4.2 Results

3.4.2.1 Simple Binding

Figure 3.12a shows a histogram comparing the distribution of measured colloid diameters obtained from two different solutions: one containing only the streptavidin and the reference colloids, and one containing both types of colloids *and* 0.1 mg/mL of monoclonal mouse anti-streptavidin antibody. As shown, there is a clear increase of 9 nm in the diameter of the streptavidin colloids in the solution containing the antibody (Figure 3.12b). We attribute this increase to the volume added to the colloid upon the specific binding to the anti-streptavidin. Specificity is demonstrated by the much smaller increase in diameter ($\sim$2.5 nm) when mixing the colloids with 0.1 mg/mL of a monoclonal isotype matched irrelevant antibody (mouse anti-rabbit; Figure 3.12b). This smaller increase is a result of non-specific binding of the irrelevant antibody to the colloids.

In Figure 3.13, we show the measured change in colloid diameter as the concentration of the specific, high-affinity antibody (monoclonal anti-streptavidin) is varied from 0.1 µg/mL to 100 µg/mL. As shown, the colloid diameter reaches its maximum value when the colloids are mixed with ≥ 5 µg/mL of antibody. Using a Bradford protein assay,[45] we determined the minimum saturating concentration of antibody for the colloid concentration in our experiment (1.2×10^9 particles/mL) to be 3.5 µg/mL, which is in good agreement with the results of our electronic pore-based immunoassay. Furthermore, the manufacturer-quoted binding capacity of the colloids indicates that each colloid has approximately 9800 streptavidin molecules on its surface. If each colloid binds to an equivalent number of antibodies, the minimum saturating concentration for a solution containing 1.2×10^9 colloids/mL will be $\sim$3.0 µg/mL; again, this is in good agreement with our results. As shown in Figure 3.13, the dynamic range of our assay corresponds to antibody concentrations from 0.5 µg/mL to the saturating concentration of $\sim$5 µg/mL. By decreasing the colloid concentration, we can decrease the binding capacity of the solution, thus decreasing the saturating concentration of antibody. In this manner, we can expect the range of sensitivity of the device to decrease to antibody concentrations as low as 10–50 ng/mL.

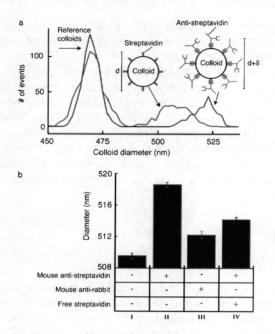

Figure 3.12 (a) A histogram showing the distribution of colloid diameters measured from a solution that contains only the reference and streptavidin colloids (green line), and a solution that contains both types of colloids and 0.1 mg/mL of monoclonal anti-streptavidin antibody (red line). The specific binding of anti-streptavidin to the streptavidin colloids produces a clear increase in the diameter of the colloids. (b) A summary of the measurements of the mean diameter of the streptavidin colloids when mixed in different solutions. A single experimental run consists of measuring several hundred colloids of each type in one solution; the plotted bars represent the mean diameter extracted from 3–5 such runs on the same solution, but using different devices. All solutions contained the streptavidin colloids and the reference colloids in a 0.5× PBS buffer (pH 7.3). The presence of additional components in each solution is indicated by a ' + ' in the column beneath the plotted bar. Column I shows the mean diameter measured without any protein added to the solution. A 9-nm increase in colloid diameter is seen in the presence of the specific antibody to streptavidin (0.1 mg/mL mouse anti-streptavidin, column II); we attribute this to the volume added to the colloid due to the specific binding of the antibody. The specificity of the probe is shown by the lack of a similar diameter increase in the presence of isotype matched irrelevant antibody (0.1 mg/mL mouse anti-rabbit, column III); the small diameter increase in this solution can be attributed to non-specific adhesion. We also perform an inhibition assay, where the specific binding of the anti-streptavidin to the colloid is disrupted by the presence of 0.2 mg/mL free streptavidin (column IV) – the presence of free antigen is shown by the decrease in diameter compared with the antigen-free solution (column II). The error bars in this figure, and in all other figures, represent the uncertainty in determining the mean diameter based on one standard deviation of the measured distributions. The dominant source of error in our measurements is the intrinsic distribution in the streptavidin colloids' diameter, with smaller contributions from the spread in diameter of the reference colloids and the electrical noise in the current measurement. From ref. 42.

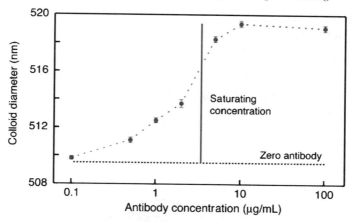

Figure 3.13   Measurements of the mean colloid diameter when mixed in solutions of
varying monoclonal mouse anti-streptavidin concentrations. The vertical
line marks the binding capacity of the colloids as determined by a
Bradford protein assay. The diameter of the colloids in the absence of
antibody is shown as the black dashed line. From ref. 42.

3.4.2.2   Inhibition Assay

We use our technique's ability to detect successfully the specific binding of
unlabeled antibodies to the colloids to perform an inhibition immunoassay. We
measure a 4.5 nm increase (see column IV of Figure 3.12b) in the diameter of
the streptavidin colloids when mixed with 0.1 mg/mL anti-streptavidin that had
been pre-incubated with 0.2 mg/mL of free streptavidin. This smaller increase
(relative to the solution containing only anti-streptavidin) indicates a decrease
in the number of antibodies binding to each colloid. We primarily attribute this
to the free streptavidin blocking the antibody binding sites. The measured
diameter of the streptavidin-coated colloid therefore indicates the presence of
free streptavidin in the solution. In general, this inhibition method can be
extended to detect any antigen that can be immobilized on the colloid surface.

The 4.5 nm increase seen in column IV of Figure 3.12b shows that some
binding of antibody to the colloid does in fact occur. Based on the control
measurement with an irrelevant antibody (column III of Figure 3.12b), we
attribute this increase to a combination of non-specific binding of blocked
antibodies, and incomplete inhibition of the antibody by the free streptavidin.
The possibility of non-specific binding does decrease the dynamic range of the
measurement. However, because of the very small uncertainty in the measured
mean colloid diameter, the dynamic range necessary to determine the amount
of ligand bound to the colloid is still quite large.

3.4.2.3   Sandwich Assay

As a second demonstration of our technique's high sensitivity to the volume
added by molecules bound to a streptavidin colloid, we perform an

immunoassay (summarized in Figure 3.14) using a sandwich configuration. Here, a primary antibody that is immobilized on the colloid surface binds to a free antigen, which in turn is bound to a secondary antibody. We immobilize the primary antibody by mixing streptavidin colloids with a biotinylated antibody (Rabbit anti-*Streptococcus* Group A) to thus create a colloid-antibody conjugate through the streptavidin-biotin bond. As shown in Figure 3.14, the measured conjugated colloids are 514 nm in diameter, a 5 nm increase over the "bare" streptavidin colloids. Next, we mix the colloid-antibody conjugates with both the specific antigen to the primary antibody (extract from a culture of *Streptococcus*

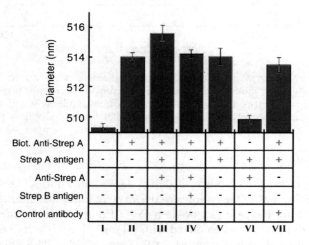

	I	II	III	IV	V	VI	VII
Biot. Anti-Strep A	-	+	+	+	+	-	+
Strep A antigen	-	-	+	-	+	+	+
Anti-Strep A	-	-	+	+	-	+	-
Strep B antigen	-	-	-	+	-	-	-
Control antibody	-	-	-	-	-	-	+

Figure 3.14 Summary of the mean colloid diameters measured when forming an antibody-antigen-antibody "sandwich" on the colloid surface. All solutions contain the reference and streptavidin colloids in a 0.5× PBS buffer (pH 7.3), along with additional components as indicated by the '+' in the column below the plotted bar. Column I indicates the measured diameter of the "bare" streptavidin colloid. We measure a ∼5 nm increase (column II) in diameter after conjugating a biotinylated antibody (biotinylated anti-*Streptococcus* Group A) to the streptavidin coated colloids. A further increase of ∼1.6 nm is seen (column III) when adding both extract from a culture of *Streptococcus* Group A and a secondary antibody specific to that antigen (unlabeled anti-*Streptococcus* Group A); this increase indicates the formation of the sandwich on the colloid surface. The specificity of the configuration is shown by the lack of an increase in diameter when adding extract from a culture of *Streptococcus* Group B (which is not bound by either antibody) in place of the Group A extract (column IV), or an irrelevant antibody in place of the specific secondary antibody (column VII). When adding the specific antigen and secondary antibody to unconjugated colloids (column VI), we measure no significant diameter increase, indicating that non-specific adhesion of antigen-secondary antibody complexes are not the cause of the diameter increase seen in column III. Finally, when adding the specific antigen alone to the conjugated colloids (column V), we see no increase in diameter, indicating that the diameter increase in column III is primarily due to the binding of the secondary antibody. From ref. 42.

Group A), and 0.1 mg/mL of a secondary antibody (unlabeled rabbit anti-*Streptococcus* Group A). Measurements of this solution show the colloids further increase in diameter by 1.6 nm. This 1.6 nm increase is not seen when the colloids are mixed with the antigen alone, indicating that the binding of the secondary antibody is the principal reason for the diameter increase. The specificity of this arrangement is demonstrated by the absence of a diameter increase in the control measurements we perform in which either the antigen or the secondary antibody is replaced by non-specific counterparts (see Figure 3.14).

It is intriguing that the measured 5 nm increase after attachment of the biotinylated antibody is less then the maximum 9 nm increase seen when utilizing the antibody-antigen bond (Figures 3.12 and 3.14) to attach antibody to the colloid. This surprising difference is most likely due to the differing conformations of the antibody in each case; however, further work is needed to clarify this. Nonetheless, despite the smaller size increase, the ability of the device to perform the sandwich assay is still clearly demonstrated.

While we have used an antibody/antigen reaction to demonstrate the power of our technique, we emphasize that its true strength is its generality: it does *not* rely on any functional properties of the free ligand. Thus, it can be applied to any ligand/receptor pair, provided the free ligand is large enough to produce a discernible change in the size of the colloid.

Future work on the device should focus on optimizing its sensitivity in terms of both ligand size (mass) and concentration. The sensitivity is dependent on four factors: the amount of ligand bound to each colloid, the intrinsic dispersion in colloid size, the colloid geometry and the colloid concentration. First, increasing the number of binding sites will lead to more ligands bound per colloid, and consequently a larger change in size. For the colloids used here, the parking area for each binding site is $\sim 80 \, nm^2$; while this is close to the steric limit for antibody molecules, the use of a smaller ligand would permit more binding sites per colloid. Second, the intrinsic spread in the sizes of the streptavidin colloids is the largest source of error in our measurement. The device's sensitivity would be enhanced by using a more monodisperse population of colloids (one with a coefficient of variation in diameter of less than 2%), or even a solution of highly monodisperse nanocrystals.[46] Third, at constant binding density, the measured change in pulse height upon binding to free ligand is proportional to the surface-to-volume ratio of the colloid. Thus, we could increase the sensitivity and dynamic range of the assay by employing a smaller colloid. For example, we estimate that using a colloid 250 nm in diameter would increase the sensitivity of the assay by a factor of four in either ligand size or concentration. Thus, based on the data shown in Figure 3.13, using a 250 nm colloid at the same particle concentration employed in this work would make the assay sensitive to either 38 kDa ligand molecules at concentrations of 0.5 μg/mL, or antibody concentrations near 0.1 μg/mL. We mention that an even more effective strategy to increase the surface-to-volume ratio would be to use a non-spherical or porous colloid (assuming the pore size is large enough to admit the free ligand) as the substrate for the immobilized receptor. Fourth, as previously mentioned, decreasing the concentration of colloids would further increase the sensitivity since it would decrease the minimum saturating concentration of free ligand. Overall, a combination of these four strategies should result in the increased sensitivity of our assay to ligand concentrations at or below 1 ng/mL.

3.5 Single-molecule Detection

To test the limits of our artificial pore device, we measured single molecules of unlabeled lambda (λ) phage DNA.[47] In more detail, we measured solutions 2.5 µg/mL λ-phage DNA in a 0.1 M KCl, 2 mM Tris (pH 8.4) buffer. The λ-phage DNA molecules were driven through the pore electrophoretically. Typical traces of the measured current are shown in Figure 3.15. The striking downward peaks, of height 10–30 pA and width 2–10 ms, correspond to individual molecules of DNA passing through the pore. In contrast, such peaks are absent when measuring only buffer. We further note that peaks are present only when using pores with diameters of 300 nm or less.

As demonstrated in Sections 3.3 and 3.4, for particles of diameter much smaller than that of the pore, the ratio of peak height to baseline current is approximately equal to the volume ratio of particle to pore: $\delta I/I \sim V_{particle}/V_{pore}$. We can estimate the volume of a single λ-phage DNA by approximating it as a cylinder with a 2-nm radius (which includes a 1-nm ionic, or Debye, layer) and a height equal to the contour length of the molecule (~ 16 µm).

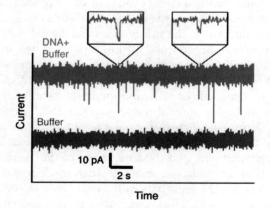

Figure 3.15 Typical traces of current *vs.* time for solutions of buffer (lower trace), and buffer with lambda phage DNA molecules (upper trace), when 0.4 V is applied across the pore. The traces are offset for clarity; the total current in each case is ~ 15 nA. Each downward spike in the lower trace represents a DNA molecule passing through the pore. The spikes are typically 2–10 ms in duration, and are well resolved, as shown in the insets. The variations in peak height most likely correspond to the different conformation of each molecule. From ref. 47.

Given the known pore volume and a total current of $I = 15$ nA, we can expect a decrease in current of $\delta I \sim 30$ pA when a DNA molecule fully inhabits the pore. This estimate agrees well with the upper range of measured peak heights. Further corroboration for this model comes from the fact that no peaks are observed when using larger pores (pores > 300 nm in diameter). When a molecule inhabits a pore with a diameter > 30 nm, the expected response in current is less than 40% of that for a 200-nm-diameter pore. Therefore, at 15 nA of total current, the maximum peak heights for a λ-phage DNA molecule will be less than 12 pA, a value not well resolvable above the noise. Our results suggest that the measured variation in δI is most likely due to differences in molecular conformation: maximum peak heights arise when an entire molecule inhabits the pore, whereas smaller peak heights occur when only a portion of a molecule resides within the pore. Future experiments should focus on controlling the conformation of each molecule to relate the measured peak height to the length of each DNA molecule. Thus, our device may provide a simple and quick method for the coarse sizing of large DNA molecules.

The results described here represent the first step toward a host of single-molecule sensing applications. Decreasing the pore size will allow us to detect and smaller-sized molecules such as proteins or viruses. The minimum achievable pore diameter for the PDMS used here is ~ 150 nm, but recent work has shown that other PDMS formulations can maintain features as small as 80 nm.[48] Finally, we can add chemical specificity in two ways: First, by covalently attaching molecules of interest to the pore wall, we expect to see changes in the transit times of molecules in solution that interact with the immobilized molecules. Second, we can measure changes in the diameter of chemically functionalized colloids upon binding of molecules in the solution, as we have already done using our immunoassays (Section 3.4).

3.6   Conclusions

We have demonstrated a novel platform, based on an on-chip artificial pore, for performing RPS measurements. Our platform capitalizes on micro- and nano-fabrication techniques that allow us to construct devices rapidly and with great flexibility in design. We have shown two applications to our pore: a label-free immunoassay and coarse-sizing of single molecules of DNA. With functionalization of the pore, we can achieve chemical specificity,[49] thus opening up new application directions in the near future.

Acronyms

RIE   reactive-ion etching
PDMS   polydimethylsiloxane
PL   photolithography
EBL   electron-beam lithography
BSA   bovine serum albumin

References

1. W. Coulter, USPTO, 1953.
2. H. Kubitschek, *Nature*, 1958, **182**, 234–235.
3. E. Gregg and K. Steidley, *Biophys. J.*, 1965, **5**, 393–405.
4. R. W. Deblois and C. P. Bean, *Review of Scientific Instruments*, 1970, **41**, 909–916.
5. R. W. Deblois, *Biophys. J.*, 1978, A149.
6. R. W. Deblois, C. P. Bean and R. K. Wesley, *J. Colloid Interface Sci.*, 1977, **61**, 323–335.
7. R. W. Deblois, E. E. Uzgiris, D. H. Cluxton and H. M. Mazzone, *Anal. Biochem.*, 1978, **90**, 273–288.
8. B. Feuer, E. E. Uzgiris, R. W. Deblois, D. H. Cluxton and J. Lenard, *Virology*, 1978, **90**, 156–161.
9. G. von Schulthess, G. Benedek and R. Deblois, *Macromolecules*, 1980, **13**, 939–945.
10. G. von Schulthess, G. Benedek and R. Deblois, *Macromolecules*, 1983, **16**, 434–440.
11. S. M. Bezrukov, I. Vodyanoy and V. A. Parsegian, *Nature*, 1994, **370**, 279–281.
12. S. M. Bezrukov, *J. Membr. Biol.*, 2000, **174**, 1–13.
13. L.-Q. Gu, O. Braha, S. Conlan, S. Cheley and H. Bayley, *Nature*, 1999, **398**, 686–690.
14. H. Bayley and C. Martin, *Chem. Rev.*, 2000, **100**, 2575–2594.
15. H. Bayley and P. Cremer, *Nature*, 2001, **413**, 226–230.
16. O. Braha, B. Walker, S. Cheley, J. Kasianowicz, L. Song, J. Gouaux and H. Bayley, *Chem. Biol.*, 1997, **4**, 497–505.
17. M. Akeson, D. Branton, J. Kasianowicz, E. Brandin and D. Deamer, *Biophys. J.*, 1999, **77**, 3227–3233.
18. J. Kasianowicz, E. Brandin, D. Branton and D. Deamer, *Proc. Natl. Acad. Sci. USA*, 1996, **93**, 13770–13773.
19. L. Sun and R. Crooks, *Langmuir*, 1999, **15**, 738–741.
20. S. B. Lee, D. T. Mitchell, L. Trofin, T. K. Nevanen, H. Soderlund and C. R. Martin, *Science*, 2002, **296**, 2198–2200.
21. L. Sun and R. Crooks, *J. Am. Chem. Soc.*, 2000, **122**, 12340–12345.
22. J. Li, D. Stein, C. McMullan, D. Branton, M. J. Aziz and J. A. Golovchenko, *Nature*, 2001, **412**, 166–169.
23. J. Li, M. Gershow, D. Stein, E. Brandin and J. Golovchenko, *Nat. Mater.*, 2003, **2**, 611–615.
24. S. M. Bezrukov and J. J. Kasianowicz, *Phys. Rev. Lett.*, 1993, **70**, 2352–2355.
25. J. W. Robertson, C. G. Rodrigues, V. M. Stanford, K. A. Rubinson, O. V. Krasilnikov and J. J. Kasianowicz, *Proc. Natl. Acad. Sci. USA*, 2007, **104**, 8207–8211.
26. A. J. Storm, C. Storm, J. Chen, H. Zandbergen, J.-F. Joanny and C. Dekker, *Nano Letters*, 2005, **5**, 1193–1197.

27. E. H. Trepagnier, A. Radenovic, D. Sivak, P. Geissler and J. Liphardt, *Nano Letters*, 2007, **7**, 2824–2830.
28. M. Wanunu and A. Meller, *Nano Letters*, 2007, **7**, 1580–1585.
29. B. Hornblower, A. Coombs, R. D. Whitaker, A. Kolomeisky, S. J. Picone, A. Meller and M. Akeson, *Nature Methods*, 2007, **4**, 315–317.
30. Q. Zhao, G. Sigalov, V. Dimitrov, B. Dorvel, U. Mirsaidov, S. Sligar, A. Aksimentiev and G. Timp, *Nano Letters*, 2007, **7**, 1680–1685.
31. Y. Xia and G. M. Whitesides, *Ange. Chem. Int. Ed.*, 1998, **37**, 550–575.
32. D. W. Fakes, M. C. Davies, A. Browns and J. M. Newton, *Surf. Interface Anal.*, 1988, **13**, 233–236.
33. M. Chaudhury and G. Whitesides, *Langmuir*, 1991, **7**, 1013–1025.
34. O. A. Saleh and L. L. Sohn, *Rev. Sci. Instrum.*, 2001, **72**, 4449–4451.
35. P. Luppa, L. Sokoll and D. Chan, *Clin. Chim. Acta*, 2001, **314**, 1–26.
36. T. Vo-Dinh and B. Cullum, *Fresenius J. Anal. Chem.*, 2000, **366**, 540–551.
37. A. Turner, *Science*, 2000, **290**, 1315–1317.
38. R. Stefan, J. van Staden and H. Aboul-Enein, *Fresenius J. Anal. Chem.*, 2000, **366**, 659–668.
39. T. Ngo, *Methods*, 2000, **22**, 1–3.
40. Y. Sykulev, D. Sherman, R. Cohen and H. Eisen, *Proc. Natl. Acad. Sci. USA*, 1992, **89**, 4703–4707.
41. W. Mullett, E. Lai and J. Yeung, *Methods*, 2000, **22**, 77–91.
42. O. A. Saleh and L. L. Sohn, *Proc. Natl. Acad. Sci. USA*, 2003, **100**, 820–824.
43. J. Anderson and J. Quinn, *Rev. Sci. Instrum.*, 1971, **42**, 1257–1258.
44. W. Smythe, *Phys. Fluids*, 1964, **7**, 633–638.
45. M. Bradford, *Anal. Biochem.*, 1976, **72**, 248–254.
46. M. Bruchez, M. Moronne, P. Gin, S. Weiss and A. Alivisatos, *Science*, 1998, **281**, 2013–2016.
47. O. A. Saleh and L. L. Sohn, *Nano Letters*, 2003, **3**, 37–38.
48. H. Schmid and B. Michel, *Macromolecules*, 2000, **33**, 3042–3049.
49. A. Carbonaro, S. Mohanty, H. Huang, L. Godley and L. L. Sohn, in press.

CHAPTER 4

Micro- and Nanocantilever Systems for Molecular Analysis

SIBANI LISA BISWAL

Department of Chemical and Biomolecular Engineering, Rice University, MS 362, 6100 Main Street, Houston, TX 77005, USA

4.1 Introduction

Sensors based on microelectromechanical systems (MEMS) have been increasingly used as mechanical transducers due to their small size, low power consumption, high sensitivity and robustness in harsh environments.[1,2] Both chemical and biological sensors usually consist of a sensitive layer or coating and a transducer. Upon interaction with a chemical species (absorption, chemical reaction, charge transfer), the physicochemical properties of the coating, such as its mass, volume, optical properties or resistance, *etc.* reversibly change.[3] These changes in the sensitive layer are detected by the transducer and translated into an electrical signal such as frequency, current or voltage, which is then read out and subjected to further data treatment and processing.

The basis of most chemical and biological sensors is that the transducer is coated with a chemically sensitive layer that will react with the target molecules. Gases such as hydrogen, carbon monoxide and nitrogen oxides react with electron-conducting oxides like tin dioxide. Oxygen, nitrogen oxide and ammonia sensors react with ion-conducting oxides like zirconium dioxide.[4] Organic layers mostly consisting of conducting or non-conducting polymers and self-assembling monolayers such as polysiloxanes, polyurethanes or polyaniline are used to sense hydrocarbons, halogenated compounds and other kinds of toxic volatile organics.[5–7]

Nano and Microsensors for Chemical and Biological Terrorism Surveillance
Edited by Jeffrey B.-H. Tok
© Royal Society of Chemistry, 2008
Published by the Royal Society of Chemistry, www.rsc.org

There are two main criteria required for a successful sensor: sensitivity and selectivity. Sensitivity characterizes the response of the sensor to a target molecule. Highly sensitive sensors provide a large signal to a small concentration of target molecules. A sensor's sensitivity increases as its surface area-to-volume ratio increases. It is for this reason that MEMS and NEMS technologies have been successful in fabricating devices that can detect target NEMS (nanoelectromechanical systems) molecules in the parts per million and the ultra-low parts per trillion range.[8] The second criterion used to characterize these sensors is selectivity, which refers to the degree of specificity to a particular target. Much of the research in sensors has been towards finding materials to prevent non-specific binding and false signals. Other figures of merit such as response time, signal stability, sensor cost and sensor reusability are all used to design a successful chemical or biosensor.

For the past decade, microcantilever devices have gained attention in the area of biological and chemical sensing applications. The invention of the atomic force microscope (AFM) in 1986 created a common tool for sending and actuating at the nanometer scale.[9] Cantilever systems have been proposed for national security and defense applications as sensor capable of detecting the presence of chemical and biological agents or explosive vapors.[10] Other applications include the use of cantilevers capable of monitoring the environment for compounds such as mercury,[11–13] or as biological detectors capable of identifying the presence of certain proteins.[11,14–17]

Most of the cantilever sensors operate on the principle that intermolecular forces that result from molecular recognition events on the surface of a cantilever produce nanoscale motion. This recognition is sensitive to picomolar concentrations.[18,19] One advantage of microcantilever sensors is that they offer real-time measurements of molecular interactions without the need for modifying the molecules of interest with external labels. Additionally, a high degree of parallelization of cantilever sensors allows for high-throughput screening as shown in Figure 4.1. There are several excellent review articles on microcantilever principles and practices.[20–22] This chapter will specifically focus this chapter on the use of nano- and microcantilevers for disease detection, calorimetry and volatile organic detection.

4.2 Dynamic Cantilever Measurements

Dynamic measurements refer to mass and surface stress changes that result in a change in the resonant frequency of the cantilever. For dynamic operation these devices are typically composed of mechanical resonators that are coated with a chemical layer such that a substance of interest will adhere to the surface of the resonator, resulting in a small change in resonator mass, thereby causing a small change in natural frequency as shown in Figure 4.2. The detection of these frequency shifts signals the presence of the target analyte. Small spring constants make it possible for measurements at high frequencies.

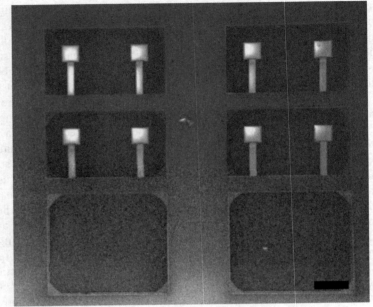

200μm

Figure 4.1 SEM of microcantilever array.

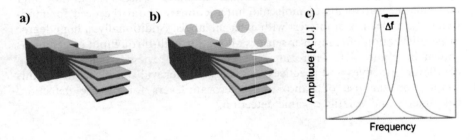

Figure 4.2 (a) Schematic of natural frequency of cantilever in dynamic mode. (b) Once mass is added to the cantilever, there is change in its resonance. (c) The resonance frequency changes can be monitored and related to mass adsorbed on the cantilever.

For a dynamic cantilever, the fundamental natural frequency is given by[23]

$$\omega_o = \sqrt{\frac{k}{m}}, \tag{4.1}$$

where k and m are the resonator's modal stiffness and modal mass, respectively. The change in frequency for a given change in mass is given by

$$\Delta\omega_o = \frac{\partial\omega_o}{\partial m}\Delta m, \tag{4.2}$$

and the resulting mass responsivity (assuming no uncertainty in system parameters, perfect measurement capabilities and very small damping) is given by

$$\Delta m = -\frac{2m}{\omega_o}\Delta\omega_o, \tag{4.3}$$

where $\Delta\omega_o$ is the measured frequency change.

The main disadvantage of dynamic sensors is that each cantilever requires a separate readout for each quantity being sensed. Devices designed for the detection of multiple, say N substances currently require N resonators, each of which must be separately sensed.

Liquid measurements are often not as sensitive in dynamic mode because the fluid offers viscous damping which decreases force sensitivity and frequency resolution.[24] For this reason, dynamic adsorption measurements are usually done in gaseous or vacuum environments.

To alleviate the drag effect in liquids, Manalis and co-workers have developed a hollow microcantilever in which fluids are passed through the cantilever.[24,25] The cantilever vibrates at its resonant frequency and when target molecules attach to the interior of the cantilever, the frequency of the cantilever changes and the mass of the molecules can be calculated by measuring in the same fashion of molecules attaching to the exterior of the cantilever in the same fashion. This hollow cantilever can be placed in a vacuum environment to further reduce its resistance to vibration. The researchers have recently measured the mass density of cells by varying the density of the surrounding solution.[25]

It is important to note that there are other factors that affect the resonant frequency as well. In fact, in addition to mass, microcantilever resonators are sensitive to viscosity, charge, and the mechanical characteristics of the chemically selective film immobilized on the device surface. For this reason, a reference sensor is used and it samples essentially the same environment as the active sensor. Therefore the charge, and viscosity effects during analyte sampling will be roughly identical on the two sensors so that the difference in resonant frequencies between the two sensors will then be due strictly to the impact of the molecular recognition event on the active sensor.

One of the goals of dynamic cantilever measurements is the ability to measure single molecules which requires sensitivity in the zeptogram (10^{-21} g) regime. The sensitivity of dynamic cantilevers depends on the noise sources and the damping of the cantilever. Noise from actuation and readout circuitry can also limit the sensitivity of the microcantilever. The minimum detectable mass is given by the ratio between the mass of the cantilever and the resonant frequency of the cantilever. The resonant frequency increases when the dimensions are decreased, so a decrease in the dimensions of the cantilever will enhance its sensitivity to mass. Nano electrical mechanical systems (NEMS) is an area of research that combines the physics of the mechanical and electrical domain at the nanometer scale. Nanocantilevers are now providing extremely high sensitivity. Recently nanocantilevers capable of detecting masses in the atto- and zeptogram $(10^{-18}-10^{-21}\,\mathrm{g})$[22,26] range have been reported. There are currently no other mass sensing techniques that can claim better sensitivity. The quartz crystal microbalance technique (QCM) has typical sensitivities in the nano-/picogram range for a single device.[27] Further discussion on nanocantilevers will be presented later in this chapter.

4.3   Static Cantilever Measurements

Surface stress changes may occur due to the difference in free energy between two surfaces of the microcantilever. In the static cantilever mode, the signal is created by a bending of the cantilever. This bending is often attributed to a surface stress caused by a chemical or biochemical reaction on the cantilever surface or by the bimorph effect caused by a change in temperature. One surface of the cantilever beam is rendered sensitive to a specific target molecule of interest, while the opposing surface is chemically passivated. The difference in surface stress induced on the sensitive relative to the passive surface of the cantilever results in a measurable mechanical deflection as demonstrated by Figure 4.3. When these target molecules interact with the sensitized surface of the cantilever, a surface stress can be induced. Additionally, surface processes like adsorption/desorption of molecules and surface reorganization induce a surface stress which causes the cantilever to be either compressive or tensile. Since the biochemical signals are translated into a mechanical signal, one can follow surface processes by measuring the bending of the

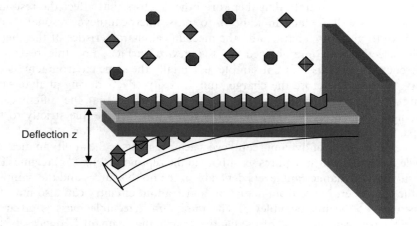

Figure 4.3   Schematic of static cantilever deflection.

cantilever. Stoney's formula relates the change in surface stress to the change in cantilever deflection.[21,28]

$$\Delta h = \frac{3\Delta\sigma(1 - \nu_1)L^2}{E_1 t_1^2} \tag{4.4}$$

where $\Delta\sigma$ is the change in surface stress, ν_1 is Poisson's ratio for the thick layer, E_1 is Young's modulus for the thick layer, t_1 is the thickness of the thick layer, L is the length of the cantilever and Δh is the deflection of the cantilever. Equation 4.4 shows a linear relation between cantilever bending and differential surface stress. For a silicon nitride cantilever 200 μm long and 0.5 μm thick, with $E = 3.20 \times 10^{11}\,N/m^2$ and $\upsilon = 0.27$, a surface stress of 0.9 mJ/m² will result in a deflection of 1 nm at the end, which can be easily detected using an optical readout system similar to that of an AFM. Stoney's formula is valid for thin films of uniform thickness. There have been several papers that describe modifications to Stoney's formula to more accurately describe the cantilever deflection to surface stress.

Most sensors utilize a cantilever coated with gold on one of its sides. Thiolated molecules can be covalently immobilized through their thiol group as they self-assemble into a monolayer onto the cantilever. The molecules can generate either a tensile stress or a compressive stress depending on whether there are attractive or repulsive interactions between the molecules. This causes the cantilever to bend respectively upward and downward if this layer is on the top side of the cantilever.

There have been a few researchers who have tackled the question of how surface stress changes cause cantilever bending. Hagan and coworkers examined the grafting of DNA strands onto the cantilever surface by balancing the reaction-induced free energy reduction on one side of the cantilever with the increase in strain energy needed so that the equilibrium of the free energy of the whole system reaches the minimum. The penalty of increasing the strain energy must be compensated for by a larger reduction in free energy due to the reaction. The four factors which were considered in Hagan's work affecting free energy reduction include electrostatic repulsions between DNA strands, osmotic pressure generated by positive counter-ions that coordinate the negatively charged DNA backbone, interactions with water, and reduction of DNA conformational entropy. Osmotic pressure and interactions of DNA with water were determined to be the most dominating factors in affecting the free energy of the system. The main disadvantage of surface-stress sensing devices is that these sensors suffer from the inability to respond to forces that vary rapidly in time.

4.4 Detection Methods

The most widely used method to detect the (static or dynamic) deflection of cantilevers is based on an optical principle, as used in the AFM. This optical lever technique employs a laser with a position-sensitive photodiode (PSD). The laser light is aimed at the tip of the cantilever so that the reflected laser light is captured upon the photodiode. The laser energy is translated into a voltage signal which allows the microcantilever displacement to be determined. This measurement method is extremely sensitive, but it requires a light reflecting surface on the cantilever and a minimum reflecting area, and thus it loses efficiency for cantilevers narrower than approximately 5 μm.

The optical detection technique is useful for measuring arrays of cantilevers simultaneously. Yue and co-workers monitored a 2-D array of hundreds of cantilevers by illuminating the entire array with a single collimated beam and capturing the reflections onto the image plane of a charge-coupled device camera. The cantilevers were fabricated with a rigid paddle on one side to provide some support in creating a flat reflecting mirror in which cantilever deflections could be captured. The need for a laser source and a detector makes it difficult to miniaturize the system, which would pose a problem for the development of portable sensor systems. The technique is ineffective for samples that absorb or scatter light (Figure 4.4).

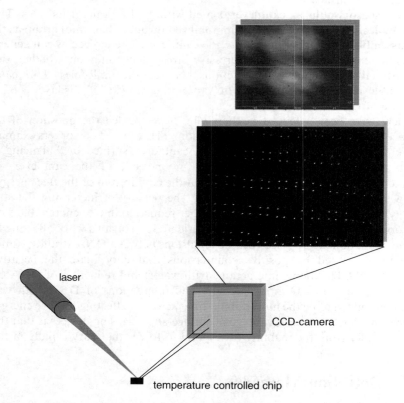

Figure 4.4   Illustration of tracking the deflection of individual cantilevers in a 2-D array simultaneously by illuminating the entire array with an expanded laser beam and capturing the cantilever reflections on a CCD camera.

Another optical method used to measure cantilever deflection is inter-ferometry. This technique uses the interference between a reference cantilever with the one reflected by the cantilever. Savran and coworkers (2003) developed a sensor comprised of two adjacent cantilevers with interdigitated fingers between them that allow interferometric detection of the relative bending. Although interferometry is highly sensitive, it does not work well in liquids.

The piezoresistive sensing has been widely used in MEMS such as pressure sensors and accelerometers.[34] Piezoresistive sensors were first applied to AFM cantilevers by Tortonese and coworkers in 1993 and have since been adopted by a number of researchers. Piezoresistive materials alter their resistance when strained. This effect is especially strong in semiconductors such as silicon. To detect the deflection of a cantilever, a resistor must be located on one of its surfaces, where the mechanical stress is maximum. Though the resolution detection is not as sensitive as the optical lever technique, its advantage is that it needs no optical components or alignment of a laser beam.

4.5 Applications of Microcantilever Sensors

4.5.1 Disease Diagnosis

The development of a tool for specific detection of biomolecular bindings such as nucleic acid and protein interactions has been sought after for disease diagnosis, drug discovery, proteomic and genomic investigation and biotoxin detection. Furthermore, a multiplexed tool capable of carrying out multiple simultaneous recognitions can be utilized for high-throughput screening. Butt and coworkers pioneered the field of surface-stress sensor devices by applying cantilever sensors to detect chemical reactions. Since this original work, surface-stress sensor devices have been applied to antibody-antigen binding. Raiteri and coworkers studied the binding of a monoclonal antibody to 2,4-dichlorophenoxyacetic acid (a herbicide) and later studied a number of other antibody-antigen binding reactions including biotin-streptavidin. Berger and coworkers detected the self-assembly of alkanethiols on the surface of a single gold-coated silicon cantilever, showing that the kinetics of this reaction obey the Langmuir isotherm kinetics.[14] Fritz and coworkers con-structed an array of eight cantilevers, each of which could be individually func-tionalized by first immersing each cantilever into unique microcapillaries, and optically interrogated using low-power laser beams and DNA immobilization and hybridization. Wu and coworkers demonstrate antigen-antibody detection of prostate specific antigen (PSA) at clinically relevant concentrations (4 ng/mL) in a background of bovine serum albumin (BSA) on single cantilevers with PSD detection (Figure 4.5). Figure 4.5 shows the steady-state cantilever deflections as a function of prostate specific antigen (PSA) concentration for three different canti-lever lengths and thicknesses. It is important to note that longer cantilevers produce larger deflections for the same PSA concentration, thereby providing higher sen-sitivity. Using 600 micron long and 0.65 micron thick silicon nitride cantilevers, it is feasible to detect a free PSA (f PSA) concentration of 0.2 ng/ml.

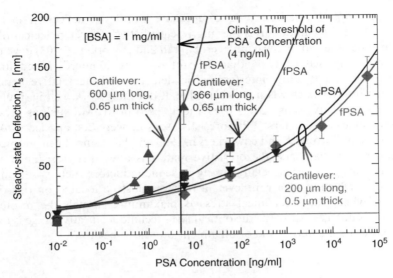

Figure 4.5 Comparison of steady-state cantilever deflections of free PSA (f PSA) and complex PSA (cPSA) at clinically relevant concentrations in a background of bovine serum albumin using three different cantilever geometries.[19]

Recently, conformational changes of surface immobilized proteins have been detected using cantilever surface-stress changes. Hegner and co-workers have immobilized a light-sensitive protein, bactriorhodopsin, onto a microcantilever and demonstrated that surface-stress changes can be detected due to protein conformational changes. Atsushi and coworkers studied the adsorption behavior of two proteins, bovine serum albumin (BSA) and Immunoglobulin G (IgG), on the surface of phase-separated organosilane monolayers cantilever sensors. The organosilane monolayers were made with noctadecyltrichlorosilane (OTS), 18-nonadecenyltriclorosilane (NTS), (2-(perfluorooctyl)ethyl)trichlorosilane (FOETS), and various mixed monolayers of these molecules were also prepared. The authors found that both BSA and IgG (at pH 7.5) were preferentially adsorbed on the FOETS phase as a single layer and as a mixture of OTS/FOETS monolayer. The authors suggested that the preferential adsorption of the proteins was due to the minimum interfacial free energy between the bulk solution and the monolayer surface as well as the electrostatic repulsion amongst charged proteins.

Groups have also used the sensitivity of cantilever mass detection for the detection of cells using the frequency shift principle.[29,30] This is certainly of great interest to the food industry and it opens the way to the detection of various bacteria. Gfeller and co-workers studied the growth of *E. coli* using an array of eight cantilevers.[31] Active sensors were coated with agarose. The authors observed no change in the resonant frequency of the reference sensors, while for the inoculated cantilevers the resonant frequency decreased exponentially over the first 5 h. The authors estimated the mass change sensitivity of the cantilever as 140 pg/Hz, from which they calculated a detected number of cells as being 200.

4.5.2 Cantilever-based Calorimetry

Cantilever-based calorimetry has been used to measure enthalpy changes in picoliter volumes of solid samples during phase transitions of n-alkanes,[32] and to investigate the thermal properties of metal clusters, for example. The layer coating the cantilever can also be catalytically active, such that heat generated directly on the surface of the cantilever due to some chemical reaction can be detected as a bimetallic deflection of the cantilever. One such example is the case of a platinum-coated cantilever, which facilitates the reaction of hydrogen and oxygen to form water. The microcantilever platform has been extended to study the thermal phase transition of biomolecules as they are heated. Microcantilever-based sensors directly translate changes in Gibbs free energy due to macromolecular interactions into mechanical responses. Majumdar and co-workers observed surface-stress changes in response to thermal dehybridization of double-stranded DNA (dsDNA) oligonucleotides that are attached onto one side of a microcantilever.[34] Once the cantilever is heated, the DNA undergoes a transition as the complementary strand melts which results in changes in the cantilever deflection. This deflection is due to changes to the electrostatic, ionic and hydration interaction forces between the remaining immobilized DNA strands. Conformational changes due to differences in the lengths and intermolecular interactions of single- and double-stranded DNA are detected as variations in cantilever deflection (Figure 4.6).

Subramanian and co-workers have designed a very sensitive glucose sensor based on the calorimetric sensitivity of microcantilevers.[35] They immobilized glucose oxidase on the gold surface of silicon nitride cantilevers and when the sensor was exposed to glucose, the cantilever bent due to the enzyme-induced exothermic processes. This method demonstrated a linear calibration curve for glucose concentrations in the range of 5–40 mM. Specificity of the sensor to glucose was shown in control experiments, with mannose at the same concentration.

4.5.3 Explosive Detection

One emerging platform for the selective detection of explosives are 'nose-on-a-chip' devices that can detect organic molecules present at concentrations as low as parts-per-billion. One example of this type of sensor is a microcantilever array in which each cantilever will be coated with a chemically sensitive material such as a polymer film designed to pick up a specific organic compound. Many of these polymer films are based on affinity absorption of certain chemicals causing the polymer film to swell. Various microcantilever platforms have been developed and used in the measurements of several chemicals and chemical properties such as gases (which includes vapor concentration, nerve agent and explosive, pH, pesticide concentration, ethanol/water concentration and ion concentration). Pinnaduwage *et al.* reported a gas sensor for the measurement of 2,4-dinitrotoluene (DNT) concentration.[36] The active

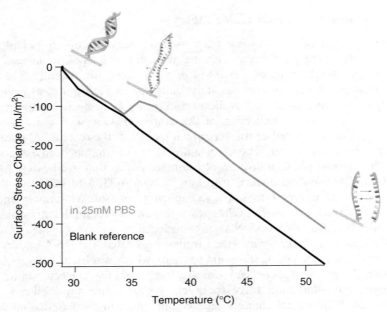

Figure 4.6   A blank cantilever shown by the black line has a linear deflection response
with respect to temperature. The cantilever with tethered DNA oligomer
displays a jump in the linear response corresponding to the melting temperature of the oligomer in a 25 mM PBS solution.[34]

sensor was coated with SXFA-[poly(1-4-hydroxy-4-trifluoromethyl-5,5,
5-trifluoro)pent-1-enyl)methylsilane] and used to sense the presence of DNT.
The authors determined the detection sensitivity of the sensor as 300 ppt. They
also showed that the adsorption process was reversible and that the detection
took place within a few seconds. In addition, they suggested that the SXFA
coat can be repeatedly exposed to varying levels of DNT concentration for over
a year and respond effectively.

Pinnaduwage *et al.* also explored the desorption characteristics of explosive
and non-explosive vapors from silicon microcantilevers without applying a
polymer coating to the cantilever.[37-39] They used deflagration of the deposited
material on the cantilever which caused thermal changes to the cantilever. The
motion of the cantilever was monitored optically using a two-quadrant posi-
tive-sensitive detector that was integrated with a spectrum analyzer for resonant frequency measurement. The explosives investigated by the authors were
trinitrotoluene (TNT), pentaerythritol tetranitrate (PETN) and hexahydro-
1,3,5-trinitroso-1,3,5-triazine (RDX) and the non-explosives were ethanol,
acetone and water vapor. TNT was found to desorb within 50 minutes in air,
while PETN and RDX took several hours to observe significant cantilever
resonant frequency response. Conversely, the kinetics of desorption for the
non-explosives was too fast to measure. The authors indicated that there is a
direct relationship between the analyte's vapor pressure and the desorbed rate.

4.6 Nanocantilevers

By using surface micromachining along with lithography techniques it is possible to create complex structures with dimensions on the order of nanometers. Recent research in the area of nanotechnology has led to the production of nanocantilevers with very high oscillation frequencies in fluid (MHz). It is important to note that these sensors are not used to sense surface stress. Instead, nanocantilevers sense analyte-induced changes that measurably alter dynamical device properties. Moreover, these structures are approaching frequencies nearing the time scale of biomolecular dynamics indicating that measurements can be taken with ever higher temporal resolution as the system mass decreases. With sufficient sensitivity, nanocantilevers are approaching the realm where single molecules may be resolvable.

Craighead and co-workers used nanocantilevers to detect a single piece of DNA 1578 base pairs in length.[40] Their work showed that they can accurately determine a molecule with mass of about 0.23 attograms (1 attogram = 10^{-18} grams). They were able to achieve such sensitivity by attaching nanoscale gold dots to the very ends of the cantilevers, which acted as capture agents for sulfide-modified double-stranded DNA. This technique with gold nanodots could be used to capture any biomolecule having a free sulfide group. The researchers' work with nanocantilevers can simplify the techniques used to screen for specific gene sequences and mutations.

Bashir and co-workers functionalized an array of nanocantilevers of[42] varying lengths with a thickness of about 30 nm and functionalized them with antibodies for viruses.[41] They found that the density of antibodies attached to the nanocantilever varied with respect to the nanocantilevers' length.

Roukes and colleagues demonstrated that their cantilevers can measure masses on the attogram scale with a resolution of just 100 zeptograms (10^{-19} grams). They placed a thin film of poly(methyl methacrylate) (PMMA) on their silicon nanocantilevers and observed a real-time response to pulses of 1,1-difluoroethane gas in air at room temperature and pressure. Mass peaks as small as 1 attogram were resolved.

The main difference between nanocantilevers and microcantilevers is the substantially reduced viscous damping when operated at atmospheric pressure allowing easier operation under ambient conditions.

4.7 Conclusions

Identification and quantitative analysis of biological and chemical molecules are vital in disease detection and monitoring, drug discovery and many sensor technologies. Compared with conventional sensors, cantilever sensors offer improved versatility. Cantilevers can operate in vacuum, gaseous and liquid environments. Static deformation and resonance frequency changes, which can be measured simultaneously, provide complementary information about the interactions between the transducers and the environment. Also, we can arrange individual cantilever transducers into large multisensor arrays for the rapid assaying of such chemical and biomolecular systems. Surface machining now allows for the fabrication of mechanical objects with lateral dimensions reaching 20 nm. These micro- and nanoscale cantilever structures are of great interest for the assaying of biomolecular masses with high sensitivity. In addition, these devices can easily be implemented into large arrays, enabling the realization of multiplexed binding assays that could identify and quantify complex mixtures with high throughput.

Acknowledgments

The author would like to thank Professor Arun Majumdar and his research group at UC Berkeley; in particular, Digvijay Roarane and Dr. Shawn Lim for their helpful discussion and contributions.

Acronyms

AFM	atomic force microscope
BSA	bovine serum albumin
DNA	deoxyribonucleic acid
DNT	2,4-dinitrotoluene
E. coli	*Escherichia coli*
FOETS	(2-(perfluorooctyl)ethyl)trichlorosilane
Ig-G	immunoglobulin G
MEMS	microelectromechanical systems
NTS	18-nonadecenyltriclorosilane
NEMS	nanoelectromechanical systems
OTS	n-octadecyltrichlorosilane
PETN	pentaerythritol tetranitrate
PMMA	poly(methyl methacrylate)
PSD	position-sensitive photodiode
PSA	prostate-specific antigen
QCM	quartz crystal microbalance
RDX	hexahydro-1,3,5-trinitroso-1,3,5-triazine
SXFA	poly(1-4-hydroxy-4-trifluoromethyl-5,5,5-trifluoro)pent-1-enyl)methylsilane
TNT	trinitrotoluene

References

1. S. S. Iqbal, M. W. Mayo, J. G. Bruno, B. V. Bronk, C. A. Batt and J. P. Chambers, A review of molecular recognition technologies for detection of biological threat agents, *Biosensors & Bioelectronics*, 2000, **15**, 549–578.
2. P. S. Waggoner and H. G. Craighead, Micro- and nanomechanical sensors for environmental, chemical, and biological detection, *Lab on a Chip*, 2007, **7**, 1238–1255.

3. D. Diamond, *Principles of Chemical and Biological Sensors*, John Wiley & Sons, 2000.
4. A. Hierlemann, O. Brand, C. Hagleitner and H. Baltes, Microfabrication techniques for chemical/biosensors, *Proceedings of the Ieee*, 2003, **91**, 839–863.
5. T. Ahuja, I. A. Mir, D. Kumar and Rajesh, Biomolecular immobilization on conducting polymers for biosensing applications, *Biomaterials*, 2007, **28**, 791–805.
6. S. Cosnier, Affinity biosensors based on electropolymerized films, *Electroanalysis*, 2005, **17**, 1701–1715.
7. P. Gouma, G. Sberveglieri, R. Dutta, J. W. Gardner and E. L. Hines, Novel materials and applications of electronic noses and tongues, *MRS Bulletin*, 2004, **29**, 697–700.
8. P. S. Waggoner and H. G. Craighead, Micro- and nanomechanical sensors for environmental, chemical, and biological detection, *Lab on a Chip*, 2007, **7**, 1238–1255.
9. G. Binnig, C. Gerber, E. Stoll, T. R. Albrecht and C. F. Quate, Atomic resolution with atomic force microscope, *Surface Science*, 1987, **189**, 1–6.
10. X. D. Yan, H. F. Ji and T. Thundat, Microcantilever (MCL) biosensing, *Current Analytical Chemistry*, 2006, **2**, 297–307.
11. S. Cherian, R. K. Gupta, B. C. Mullin and T. Thundat, Detection of heavy metal ions using protein-functionalized microcantilever sensors, *Biosensors & Bioelectronics*, 2003, **19**, 411–416.
12. B. Rogers, L. Manning, M. Jones, T. Sulchek, K. Murray, B. Beneschott, J. D. Adams, Z. Hu, T. Thundat, H. Cavazos and S. C. Minne, Mercury vapor detection with a self-sensing, resonating piezoelectric cantilever, *Review of Scientific Instruments*, 2003, **74**, 4899–4901.
13. X. H. Xu, T. G. Thundat, G. M. Brown and H. F. Ji, Detection of Hg2+ using microcantilever sensors, *Analytical Chemistry*, 2002, **74**, 3611–3615.
14. R. Mukhopadhyay, V. Sumbayev, M. Lorentzen, J. Kjems, P. A. Andreasen and F. Besenbacher, Cantilever sensor for nanomechanical detection of specific protein conformations, *Nano Letters*, 2005, **5**, 2385–2388.
15. P. Dutta, C. A. Tipple, N. V. Lavrik, P. G. Datskos, H. Hofstetter, O. Hofstetter and M. J. Sepaniak, Enantioselective sensors based on antibody-mediated nanomechanics, *Analytical Chemistry*, 2003, **75**, 2342–2348.
16. G. H. Wu, H. F. Ji, K. Hansen, T. Thundat, R. Datar, R. Cote, M. F. Hagan, A. K. Chakraborty and A. Majumdar, Bioassay of prostate-specific antigen (PSA) using microcantilevers, *Nature Biotechnology*, 2001, **19**, 856–860.
17. N. Backmann, C. Zahnd, F. Huber, A. Bietsch, A. Pluckthun, H. P. Lang, H. J. Guntherodt, M. Hegner and C. Gerber, A label-free immunosensor array using single-chain antibody fragments, *Proceedings of the National Academy of Sciences of the United States of America*, 2005, **102**, 14587–14592.
18. N. V. Lavrik and P. G. Datskos, Femtogram mass detection using photothermally actuated nanomechanical resonators, *Applied Physics Letters*, 2003, **82**, 2697–2699.

19. N. V. Lavrik, M. J. Sepaniak and P. G. Datskos, Cantilever transducers as a platform for chemical and biological sensors, *Review of Scientific Instruments*, 2004, **75**, 2229–2253.
20. T. Thundat and A. Majumdar, Microcantilevers for Physical, Chemical, and Biological Sensing in *Sensors and Sensing in Biology and Engineering*, Springer-Verlag, New York, 2003.
21. C. Ziegler, Cantilever-based biosensors, *Analytical and Bioanalytical Chemistry*, 2004, **379**, 946–959.
22. M. Li, H. X. Tang and M. L. Roukes, *Nature Nanotechnology*, 2007, **2**, 114–120.
23. A. Sebastian, A. Gannepalli and M. V. Salapaka, A review of the systems approach to the analysis of dynamic-mode atomic force microscopy, *Ieee Transactions on Control Systems Technology*, 2007, **15**, 952–959.
24. T. P. Burg and S. R. Manalis, Suspended microchannel resonators for biomolecular detection, *Applied Physics Letters*, 2003, **83**, 2698–2700.
25. M. Godin, A. K. Bryan, T. P. Burg, K. Babcock and S. R. Manalis, Measuring the mass, density, and size of particles and cells using a suspended microchannel resonator, *Applied Physics Letters*, 2007, **91**.
26. J. L. Arlett, J. R. Maloney, B. Gudlewski, M. Muluneh and M. L. Roukes, Self-sensing micro- and nanocantilevers with attonewton-scale force resolution, *Nano Letters*, 2006, **6**, 1000–1006.
27. G. S. Huang, M. T. Wang, C. W. Su, Y. S. Chen and M. Y. Hong, Picogram detection of metal ions by melanin-sensitized piezoelectric sensor, *Biosensors & Bioelectronics*, 2007, **23**, 319–325.
28. G. G. Stoney, The tension of metallic films deposited by electrolysis, *Proceedings of the Royal Society of London Series A-Containing Papers of a Mathematical and Physical Character*, 1909, **82**, 172–175.
29. D. Ramos, J. Tamayo, J. Mertens, M. Calleja, L. G. Villanueva and A. Zaballos, Detection of bacteria based on the thermomechanical noise of a nanomechanical resonator: origin of the response and detection limits, *Nanotechnology*, 2008, **19**.
30. A. P. Davila, J. Jang, A. K. Gupta, T. Walter, A. Aronson and R. Bashir, Microresonator mass sensors for detection of Bacillus anthracis Sterne spores in air and water, *Biosensors & Bioelectronics*, 2007, **22**, 3028–3035.
31. K. Y. Gfeller, N. Nugaeva and M. Hegner, Micromechanical oscillators as rapid biosensor for the detection of active growth of Escherichia coli, *Biosensors & Bioelectronics*, 2005, **21**, 528–533.
32. R. Berger, C. Gerber, J. K. Gimzewski, E. Meyer and H. J. Guntherodt, Thermal analysis using a micromechanical calorimeter, *Applied Physics Letters*, 1996, **69**, 40–42.
33. T. Bachels, F. Tiefenbacher and R. Schafer, Condensation of isolated metal clusters studied with a calorimeter, *Journal of Chemical Physics*, 1999, **110**, 10008–10015.
34. S. L. Biswal, D. Raorane, A. Chaiken, H. Birecki and A. Majumdar, Nanomechanical Detection of DNA Melting on Microcantilever Surfaces, *Analytical Chemistry*, 2006.

35. A. Subramanian, P. I. Oden, S. J. Kennel, K. B. Jacobson, R. J. Warmack, T. Thundat and M. J. Doktycz, Glucose biosensing using an enzyme-coated microcantilever, *Applied Physics Letters*, 2002, **81**, 385–387.

36. L. A. Pinnaduwage, T. Thundat, J. E. Hawk, D. L. Hedden, R. Britt, E. J. Houser, S. Stepnowski, R. A. McGill and D. Bubb, Detection of 2,4-dinitrotoluene using microcantilever sensors, *Sensors and Actuators B-Chemical*, 2004, **99**, 223–229.

37. L. A. Pinnaduwage, T. Thundat, A. Gehl, S. D. Wilson, D. L. Hedden and R. T. Lareau, Desorption characteristics, of uncoated silicon micro-cantilever surfaces for explosive and common nonexplosive vapors, *Ultramicroscopy*, 2004, **100**, 211–216.

38. L. A. Pinnaduwage, D. Yi, F. Tian, T. Thundat and R. T. Lareau, Adsorption of trinitrotoluene on uncoated silicon microcantilever surfaces, *Langmuir*, 2004, **20**, 2690–2694.

39. L. A. Pinnaduwage, L. A. Pinnaduwage, A. Wig, D. L. Hedden, A. Gehl, D. Yi, T. Thundat and R. T. Lareau, Detection of trinitrotoluene via defla-gration on a microcantilever, *Journal of Applied Physics*, 2004, **95**, 5871–5875.

40. B. Ilic, Y. Yang, K. Aubin, R. Reichenbach, S. Krylov and H. G. Craig-head, Enumeration of DNA molecules bound to a nanomechanical oscillator, *Nano Letters*, 2005, **5**, 925–929.

41. A. K. Gupta, P. R. Nair, D. Akin, M. R. Ladisch, S. Broyles, M. A. Alam and R. Bashir, Anomalous resonance in a nanomechanical biosensor, *Proceedings of the National Academy of Sciences of the United States of America*, 2006, **103**, 13362–13367.

42. Y. T. Yang, C. Callegari, X. L. Feng, K. L. Ekinci and M. L. Roukes, Zeptogram-scale nanomechanical mass sensing, *Nano Letters*, 2006, **6**, 583–586.

CHAPTER 5

Fiber-optic Sensors for Biological and Chemical Agent Detection

MATTHEW J. AERNECKE AND DAVID R. WALT

Tufts University, Department of Chemistry, 62 Talbot Ave, Medford, MA 02155, USA

5.1 Introduction

Monitoring chemical and biological warfare agents (CWAs and BWAs) continuously with adequate speed, sensitivity and specificity poses a formidable challenge. Several approaches, based on a wide range of chemical and biological detection principles, have been explored with mixed success.[1] Fiber-optic sensors provide a versatile platform for CWA and BWA detection that can be easily tailored to a specific problem and offer unique advantages, such as small size, minimal equipment requirements and high multiplexing ability, which make them a viable option for continuous threat monitoring. In this chapter, we discuss the general utilization of optical fibers as sensors, and present specific examples where this platform has been applied to BWA and CWA detection. We limit our discussion to those methods that utilize electromagnetic radiation in the visible range of the spectrum.

5.2 Fiber-optic Basics

Optical fibers or optical waveguides are dielectric structures that are capable of channeling and conducting light. They consist of a core, a cylindrical structure

Nano and Microsensors for Chemical and Biological Terrorism Surveillance
Edited by Jeffrey B.-H. Tok
© Royal Society of Chemistry, 2008
Published by the Royal Society of Chemistry, www.rsc.org

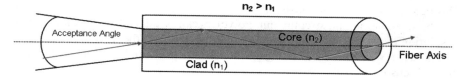

Figure 5.1   Lengthwise cross-section of a step-index optical fiber illustrating light transmission *via* Total Internal Reflection (TIR).

through which light is transmitted, and the clad, a second material that completely surrounds the core. Light transmission occurs in these structures when the refractive index of the core is higher than the refractive index of the clad. Under these conditions (Figure 5.1), light that is incident on the core and is within the acceptance angle of the fiber will penetrate into the core and eventually reflect off the core-clad interface. This reflection process, known as total internal reflection (TIR), occurs repeatedly as the ray strikes subsequent core-clad interfaces thereby trapping it in the medium of higher refractive index *i.e.* the core.

The core may be comprised of either a uniform refractive index material, known as a step-index fiber, or it may vary in composition, which is known as a graded refractive index, or GRIN, fiber. Since the majority of optical fibers used as sensors are step-index fibers, GRIN fibers will not be dealt with in any great detail here, however the reader is referred to Ref. 2 for a more detailed theoretical discussion.

Optical fibers are most commonly fabricated from heavy metal doped silica or glass; however, polymers are also widely used. The composition of the fiber determines what wavelength range it is able to transmit. Most formulations are tailored to a particular region of the electromagnetic spectrum. Those fibers designed for visible applications are capable of transmitting light across the visible spectrum and into the near IR. The transmission range for various formulations is generally around 405–700 nm for polymer fibers, 360–710 nm for glass fibers and 300–700 nm for high-quality quartz fibers. The choice of a particular fiber is dictated by the cost, quality and spectral region necessary for the application.

5.3   Optical Fibers as Sensors

Optical fibers are transformed into chemical or biological sensors by attaching a chemical recognition element to one end. When the target analyte is present, there is a change in the optical properties of the recognition element. These optical properties can be a result of changes in refractive index, reflectivity, absorbance, fluorescence or chemiluminescence. Some of the simplest modifications for attaching sensing chemistry involve physical adsorption or fastening a functionalized membrane to the fiber surface. Because these methods are prone to desorption or membrane loss, they are generally avoided in favor of a more durable coupling of the sensing chemistry. Such coupling can be accomplished through the use of an organosilane or by a high-affinity interaction such as the one between biotin and streptavidin. Polymers can be added by dip-coating from a solution or by polymerizing the material directly onto the fiber surface.

There is a wide variety of recognition elements that can be incorporated into optical sensors. Sensing chemistries can range from a fluorophore that changes its fluorescence intensity with pH, to a complex multicomponent system capitalizing on the specific recognition of biological elements such as nucleic acids, enzymes, antibodies or even whole cells. Antibody-based biological recognition elements utilized in binding assays are typically labeled with a reporter molecule, while enzyme-catalyzed detection reactions are coupled with an optical indicator sensitive to one or more of the reaction products.

The use of optical fibers as sensors has several advantages over other types of sensors. Optical fibers are able to transmit several wavelengths of light simultaneously, therefore multiple signals can be carried over a single fiber providing a platform capable of a high degree of multiplexing. Additionally, because the signals are optical, they are free from electrical or magnetic interference and can work reliably in harsh environments. A disadvantage of optical fibers is that ambient light can interfere with the analytical signal. This problem can be obviated in some systems by modulating the signal.

The sensing systems discussed herein are either absorbance or fluorescence based and, as such, contain a source of excitation energy, focusing optics, an optical fiber that has been modified with a chemical recognition element and a detector. Excitation sources have traditionally been lasers or polychromatic light sources that are attenuated through the use of a filter or monochromater. An increasing number of applications are utilizing ultra-bright LEDs as an excitation source due to their relatively narrow bandwidth and low power consumption. Typical detectors include photodiodes, diode arrays, charge coupled device (CCD) cameras and Complementary Metal-Oxide-Semiconductor (CMOS) cameras.

5.4   Biological Agent Detection

Along with the standard demands of speed and specificity, detecting biological warfare agents has the additional challenge of scope. Any approach must be tailored to detect a broad range of analytes including proteins, viruses, whole cells and spores. The magnitude of the problem becomes readily apparent when one considers that there are approximately 1700 microorganisms that can be pathogenic to humans.[1] Conventional cell culture techniques are specific and quantitative but time consuming. Alternative methods are needed if real-time monitoring of several BWAs is to be realized. The optical-fiber-based approaches discussed in this section represent a step in this direction.

5.4.1   Fiber Optic Immunosensors

The RAPTOR (Rapid Automatic and Portable Fluorometer Assay System) and its predecessor systems, the Analyte 2000 and MANTIS, all utilize a fluorescent sandwich immunoassay coupled with evanescent wave sensing on fiber-optic waveguides (Figure 5.2). These systems, developed by the US Naval Research Laboratory and private sector firms, enable direct detection of pathogens either as whole cells or spores, or the assay can target protein markers, enabling blood and other fluids to be analyzed for signs of exposure.

The sensing configuration used in these instruments takes advantage of the small electromagnetic field that extends beyond the surface of a non-clad fiber core when it is placed in a dielectric medium of lower refractive index, such as

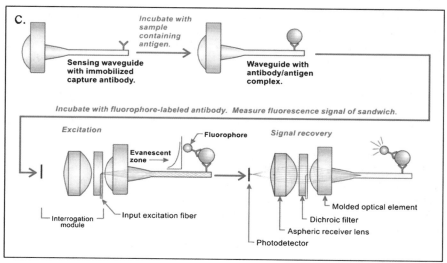

Figure 5.2 (a) Image of the RAPTOR. (b) Image of the BioHawk portable 8-channel assay system with automated sample collection capability. (c) Schematic of the sandwich immunoassay and evanescent wave detection employed in the RAPTOR. Reprinted with Permission from Research International, Inc.

an aqueous solution. The depth of penetration of this field, known as the evanescent wave, is on the same order of magnitude as the wavelength of light traveling through the fiber. In order to convert a conventional step-index fiber into an evanescent wave sensor, the cladding material is stripped and the chemical recognition element is attached directly to the exposed core surface. The sensor is then placed into solution and the analysis is performed.

In the RAPTOR, the recognition elements are target-specific capture antibodies that are covalently bound to the surface of the core. An analyte solution washes over the sensor for a period of time during which antigen-antibody binding occurs. The analyte solution is removed and a second solution containing fluorescently labeled detection antibodies is added. The detection antibodies bind to the antigen-capture antibody complex and the aggregate is excited *via* the evanescent wave. Emitted fluorescence from the complex is collected by the same fiber and transmitted back to a photodetector. Due to the limited depth of penetration of the evanescent wave, the return signal can be weak, particularly if the recognition element is large and not directly attached to the fiber surface. The analytical signal can be intensified if a longer portion of the fiber is utilized for sensing.

Early non-portable single-channel versions of this device utilized silica waveguides (200 μm diameter) that were taper-etched and coated with adsorbed capture antibodies. Using rhodamine-labeled detection antibodies, this early system exhibited limits of detection of 5 ng/ml for botulinum toxin[3] and *Y. Pestis* fraction 1 antigen.[4,5] Response times were 1 minute and less than 15 minutes, respectively.

Through several iterations this platform has evolved into a portable, multi-component system for bioagent detection. The Analyte 2000 is the first example of a multiplexed sensor that contains four independent evanescent waveguides arranged in a fluidic coupon that is connected to an external reagent delivery system. Excitation light at 635 nm is generated by a laser diode and fluorescence emission is detected with a photodiode. Using Cy5 labeled detection antibodies in a sandwich immunoassay format, this mobile autonomous instrument has demonstrated detection limits of 100 pg/ml of ricin under laboratory conditions and 1 ng/ml of ricin in river water.[6] More recently, the Analyte 2000 has been able to detect *B. anthrasis* spores directly in various powdered matrices[7] and *Vaccinia* virus in spiked throat swabs.[8]

This system has also been used in a competitive immunoassay format for multiplexed detection of the explosives TNT[9,10] and RDX.[10] In the competitive assay format, a sample solution is mixed with Cy5 labeled TNT and RDX analogues and flowed over waveguides functionalized with anti-TNT and anti-RDX capture antibodies. The labeled and non-labeled analytes compete for a limited number of binding sites on the sensor. The fluorescent response obtained from the sample solution is compared to a reference response collected from a solution containing only labeled antigen. A net decrease in the fluorescent signal from the sample solution constitutes a positive result.

This instrument was also modified for use on an unmanned aerial vehicle (UAV) and utilized in the field to detect aerosolized bacterial spores.[11] Field samples were introduced *via* an external air sampler that takes advantage of the approximately 100 L/min air flow generated during UAV flight. This air flow was directed into a rapidly circulating aqueous film on the inner walls of the sampling chamber. Aerosolized particles are concentrated in the aqueous phase and after a specified collection time, the sample solution is directed towards the instrument for analysis. The results of the fully automated assay were transmitted to the ground over a wireless datalink. The instrument performed an analysis every 5 minutes for the duration of a 15–20 minute flight. It was successful in detecting *B. globigii* spores in 4 out of 9 controlled releases. The detection limits of both the UAV and the aforementioned ricin studies were improved through the use of biotin-streptavidin coupling of the capture antibody to the optical-fiber surface.

The large size of the multi-component Analyte 2000 was reduced in a consolidated version dubbed the MANTIS (Man-Portable Analyte Identification System)[12] and further optimized to its current form, the RAPTOR (Rapid Automatic and Portable Fluorometer Assay System).[13] Upgrades to this design include onboard reagent storage and delivery along with an injection-molded four-channel polystyrene waveguide housed in an easily exchanged microfluidic coupon. The low-voltage portable system has been reduced in weight to approximately 14 lb. Optimization of the optical component assembly has increased the penetration depth of the evanescent field from the waveguide's surface to approximately 2500 nm, maximizing its coupling to captured targets. Replacing Cy5 with Alexa Fluor 647 as the detection antibody label has resulted in lower detection limits.[14] Alexa Fluor 647 has a lower tendency to self-quench, enabling higher dye-to-protein ratios which provide stronger fluorescent signals.[14] Because the RAPTOR can analyze up to four separate waveguides at a time, assays can be run for a single analyte with replicates or for multiple analytes simultaneously.[15,16] Detection limits of 5×10^4 cfu/ml for *Bacillius globigii*, 50 ng/ml for ricin and 5×10^5 cfu/ml for *F. Tularensis* have been demonstrated on a single multiplexed coupon.[15] The non-destructive nature of the sandwich assay enables confirmatory analysis by PCR amplification after culturing the cells captured by the coupon.[17,18] The shelf life of a typical coupon is 5–6 months and several different coupons can be used with a single instrument.[13] The RAPTOR represents the only completely integrated fiber-optic-based biological agent monitoring system to date.

5.4.2 Optical-fiber Nucleic Acid Sensors

A powerful aid in gauging the severity and extent of a biological attack is the ability to screen for all potential pathogens rapidly and in parallel. The advent of DNA microarray technologies has enabled this type of concurrent multi-analyte detection for a wide range of micoorganisms and, recently, Song *et al.* have adapted this technology to detect potential BWAs.[19] Their method employed a multiplexed fiber-optic DNA microarray to simultaneously detect pathogens in autoclaved cultures and spiked wastewater samples.[20]

The authors used an imaging fiber-optic bundle as the platform for their multiplexed array. An imaging fiber-optic bundle is an array of hundreds to

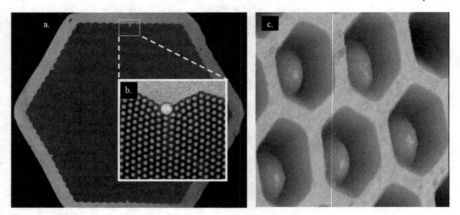

Figure 5.3 (a) Overview of a hexagonally packed imaging fiber-optic bundle. Each individual fiber is 3 microns. (b) Magnified image of (a) to show the arrangement of fibers in the bundle. (c) AFM image of etched microwells containing microbead sensors. Reprinted with permission from Ref. 54. Copyright 2003 Royal Society of Chemistry.

tens of thousands of individual optical fibers fused together in a manner such that the location of each fiber is retained at both ends (Figure 5.3). Due to their coherent nature, these bundles can transmit images from one end to the other. Additionally, each individual fiber can be modified to function as a sensor either by coating one surface of the fiber with sensing chemistry or by creating an array of microwells into which functionalized sensing microspheres of complementary size are distributed. In the latter approach, the fiber cores are preferentially etched using an acidic solution. When a suspension of microspheres is placed on the microwell array, they randomly distribute across the etched face of the fiber and assemble into the wells during evaporation, where they remain due to capillary forces. Multiple populations of microspheres with different sensing chemistries can be prepared, combined and distributed into the wells enabling a high degree of multiplexing.

The organisms targeted in these studies were *B. anthracis, Y. pestis, F. tularensis, B. melitensis, C. botulinum, Vaccinia* virus and *B. thuringiensis kurstaki*. Species-specific 50mer nucleic acid probe sequences (two per organism and six for *B. anthracis*) were coupled to the surface of 3.1-μm diameter polystyrene microspheres and distributed into microwells etched into the polished end of an imaging fiber-optic bundle. The large number of microsphere types (18 total) present in the array were encoded using two fluorescent dyes at several discrete concentrations. The sensitivity of each custom-designed probe sequence was tested by direct hybridization to a fluorescently labeled synthetic target at varying concentrations. After establishing optimal hybridization conditions, sequence specificity and verifying low array cross-reactivity, the sensor was challenged with various mixtures of the autoclaved BWAs and spiked wastewater samples. The samples were first enriched using PCR with

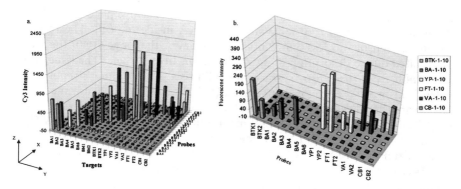

Figure 5.4   (a) Cross-reactivity of the BWA bead-based microarray. Probe strands attached to the microsphere surface are displayed on the x-axis and synthetic targets are displayed on the y-axis. (b) Multiplexed detection of autoclaved BWA samples (1 : 10 concentration) in spiked wastewater using the microarray. Reprinted with permission from Ref. 20. Copyright 2006 American Chemical Society.

Cy3-labeled primers and the amplified target sequences were detected by direct hybridization to the array. The array responded within 30 minutes to target concentrations as low as 10 fM using a minimal volume of sample. The results of these experiments are presented in Figure 5.4.

There are several advantages to using bead-based microarray technologies for the detection of BWAs. Captured target sequences can be dehybridized and the multiplexed DNA microarray can be reused with minimal signal degradation.[21] The bead-based platform makes the incorporation of additional target sequences facile, provided there is no cross-reactivity with existing beads. The high degree of redundant bead types present in the array increases signal-to-noise ratios and minimizes the likelihood of false positives and false negatives because each result is verified over multiple independent measurements. Additionally, the small size of the array (0.5 mm) enables the use of small sample volumes and can be easily integrated into microfluidic devices.

5.4.3   Sensors for Biomolecules

Monitoring methods involving PCR amplification or using multi-step protocols, such as the sandwich immunoassay, are time consuming. For rapid detection of a BWA release, a non-specific sensor can be used to detect the signatures of pathogenic microorganisms rather than identify each specific BWA. This broad detection approach has been demonstrated by Tabacco and co-workers[22–25] using nucleic acid or lipid specific fluorescent indicators that increase their quantum yield when they interact with these biomolecules. Early studies demonstrated that SYTO 13, an intercalating DNA dye, can be used to detect *Pseudomonas aeruginosa* in both aqueous and aerosolized samples.[22] A 1.4-mm single-core optical-fiber was coated with a thin film of SYTO 13 and exposed to various concentrations of bacterial cells. The dye penetrated the cell membrane and bound to double-stranded DNA inside the cell. The response time of the sensor was less than 2 minutes, however, the longevity of the sensor was limited due to desorption of the dye from the fiber surface. The issue of dye desorption from the sensor surface was addressed in subsequent reports that utilized adsorbed[23] and covalently linked[25] indicator doped dendrimer films. This format has been expanded to include other intercalating DNA dyes such as SYTOX Green,[23] as well as lipid membrane dyes such as FAST DiA.[25]

The same format has been expanded beyond whole-cell detection to include bacterial endospores, such as those formed by *B. anthracis*. The outer spore coat has been shown to contain a high concentration of calcium.[24] This calcium-rich environment can be detected specifically by using the dye calcein, which forms a strongly fluorescent complex when chelated with calcium ions. Calcein was dissolved in glycerol and coated onto a disposable planar glass coupon. The sensing coupon was optically interrogated from below with a single-core optical-fiber. When endospores came in contact with the sensing film, the dye chelated with calcium ions present in the endospore shell and generated an increase in fluorescence intensity at 475 nm. The onset of this response occurred within two minutes and reached a maximum within twenty minutes. The limit of detection for this method was 1763 spores and there was little or no cross-reaction with other biological materials such as bacteria, viruses, fungal spores or pollen.

5.5   Chemical Agent Detection

Detecting chemical warfare agents (CWAs) poses the challenge of isolating and identifying a transient species in a continuously fluctuating ambient environment. Ideally, a CWA monitoring system should respond rapidly and specifically, and should consist of a network of correlated sensors capable of tracking the movement of a hazardous plume. The most conclusive means of identifying CWAs are mass spectroscopy- (MS-) based methods. MS instruments are difficult to integrate into a widespread sensing network and, until recently, have lacked the mobility necessary for efficient functioning in the field. Fiber-optic based CWA sensors, due to their small size and ability to transmit analytical signals over long distances, offer a versatile platform for continuous monitoring both locally and over a large area.

5.5.1   Polymer-based Fiber-optic Sensors

5.5.1.1   Single-fiber Systems

Traditionally, polymers have been used in conjunction with fiber-optics primarily as a support for attaching sensing chemistry, but these materials can also function directly as the sensor itself. Work by Bansal and El-Sharif used an environmentally sensitive polymer as the detection element in a fiber-optic based dimethyl methylphosphonate sensor.[26] A 1-cm length of cladding was chemically etched from the midpoint of a meter-long single-core optical fiber. The exposed core was coated in a layer of polypyrrole doped with one of three acidic compounds. Upon exposure to a chemical agent, the transmission properties of the fiber at 633 nm decreased due to a change in the refractive index of the polymer-clad region. Optimal signal enhancement was found to be dependent upon the thickness of the polymer layer, the type of dopant used and the concentration of Cu^{+2} ions it contained. The maximum decrease in transmitted light intensity possible under optimized conditions was 25%. The sensor was also shown to cross-react with vapors such as acetone and ammonia. This type of non-selective sensing approach suffers from a poor dynamic response range as well as cross-reactivity with other vapors.

Greater specificity in the identification of nerve agent can be achieved when the signal generated by the sensor is dependent on several chemical parameters. This multi-faceted approach was used by Jenkins *et al.*, who combined the specificity of a molecularly imprinted polymer (MIP) with ligand-dependent fluorescence enhancement. The sensing molecule used in these studies was a fluorescent europium complex that incorporated polymerizable ligands and organophosphate target molecules into its coordination sphere. The monomer, vinyl or divinylmethyl benzoate, was used as a crosslinking agent for the MIP. Analyte specificity is produced by forming the polymer around the complex with the target molecule coordinated to the polymer side chains. Following polymerization, the target molecule is extracted, leaving a sensing "pocket" that is tailored to the shape and binding orientation of the molecule. Upon subsequent re-exposure of the MIP to the analyte, the fluorescence intensity of the MIP increases and, in some cases, a separate analyte-specific peak evolves.[27] The magnitude of this response can then be used to determine the concentration of the analyte. This approach has been used to fabricate sensors with a high degree of specificity for pesticides,[28] the chemical warfare agents EA2192, VX, sarin and soman,[29] and the hydrolysis products of sarin and soman.[27]

The MIP sensing element can be readily applied to the tapered end of an optical fiber *via* dip-coating with the polymer-lanthanide reaction mixture and curing under a UV light. Evanescent coupling was used to excite the fluorophore and the analyte-sensitive emission was monitored with a portable spectrometer. The sensors reached their maximum response within 8 to 15 minutes. Compounds structurally similar to chemical warfare agents exhibited minor effects on the emission spectrum of the MIP; however, these effects could be distinguished from those that occurred when the target analyte bound. A portable system was constructed using this technique that had a detection limit in water of 11 parts per trillion (ppt) EA2192, 24 ppt sarin, 33 ppt soman and 21 ppt VX.[29]

5.5.1.2 Artificial Nose Systems

Artificial or electronic noses use an array of cross-reactive semi-selective sensors that differentially interact with a vapor-phase analyte to produce a response pattern. These aggregate responses are used to train a pattern recognition program to identify subsequent exposures of the array to the learned vapors.

These systems respond reversibly to numerous types of vapors and offer a compact platform that is ideal for continuous environmental monitoring. In much the same way that biological systems can be trained to recognize new odors, artificial nose systems are theoretically limited only by the number of vapors in their training database. There are several artificial nose systems described in the literature, including those that use surface acoustic waves (SAW), carbon-black polymer chemiresistors, metal-oxide field effect transistors (MOSFETS) and quartz crystal microbalances (QCMs) as the vapor-sensitive elements.

One of the first reports that employed a fiber-optic sensor for organic vapor detection was in 1991, where the end of a single-core fiber-optic was chemically modified with a layer of polydimethylsilicone polymer and the solvatochromic fluorescent indicator Nile Red.[30] As a vapor diffused into the polymer, the microenvironment surrounding the dye molecules changed due to the polarity of the vapor and the degree of polymer swelling that occurred due to the absorption of vapor. These effects were reported optically as a change in the wavelength and/or intensity of the Nile Red fluorescence peak. This sensor was shown to respond differentially to benzene, toluene, ethyl benzene, xylene and gasoline at 100 ppm within 2.5 minutes of vapor exposure.[30] A miniaturized version of this platform was produced and successfully field tested at a site contaminated with jet fuel. This report laid the groundwork for more complex multi-component polymer-based systems employing pattern recognition algorithms.[31,32] The use of several individual single-core fibers coated with different polymer matrices and bundled together generated a diverse collection of semi-selective responses that has been shown to correctly classify a variety of organic vapors with upwards of 90% accuracy.

Effectively reproducing the polymer-sensing elements both within and between arrays is problematic and therefore inhibits the transfer of a pattern recognition classifier between arrays. Considering the average odor memory required to solve a variety of odor recognition problems can potentially contain thousands of target compounds, repeatedly training each new set of sensors quickly becomes a prohibitive task. This fundamental limitation led to the development of bead-based fiber-optic vapor sensors.[33] In this scheme, surface functionalized porous silica microspheres are coated with a vapor-sensitive dye. The dye can be either adsorbed onto the microsphere surface or covalently linked. The signals obtained from such sensors, much like their polymer counterparts, are representative of the polarity of the analyte vapor and the degree to which it interacts with the surface. Billions of identically responding microsphere sensors can be fabricated in a few simple steps and the diversity of sensor types is limited only by the types of dyes or surface functionalities available.

Multiplexed vapor sensing arrays are produced by combining sensors from several different bead stocks and distributing them randomly onto the distal end of an etched fiber-optic array as described in Section 5.4.2 above. Individual beads are identified by analyzing their intrinsic response to a predefined vapor and comparing this response to a library of reference responses from bead types stored from previous exposures (Figure 5.5).

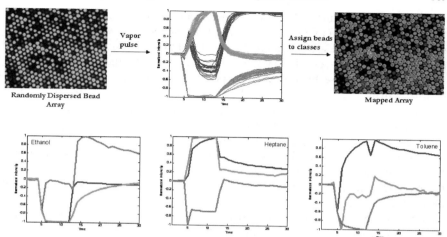

Figure 5.5   Imaging fiber-optic array vapor sensor operation. A randomized array is first decoded by analyzing its intrinsic response to a predefined vapor. The mapped array is then exposed to subsequent vapors and the response patterns are analyzed with pattern recognition software, compared to a learned database, and identified. Adapted from Ref. 33 with permission. Copyright 1999 American Chemical Society.

This method obviates the need for a separate encoding chemistry and provides a simple, rapid and low-cost means of producing arrays that respond similarly. The existing platform can be easily updated as new sensors are developed. The large degree of sensor redundancy present in each array allows a small section of the array to be imaged over several thousand vapor exposures, greatly prolonging the lifetime of each individual array.[34] As the sensor responses degrade over time due to photobleaching and sensor poisoning, they can be replaced without having to retrain the pattern recognition algorithm.[35]

Bead-based sensing arrays have been utilized in the detection of explosive vapors both in the laboratory[36] and using a portable field instrument with high success rates.[37] This sensing approach has also been used to identify the nerve agent simulant dimethyl methylphosphonate (DMMP) correctly 100% of the time when it was presented randomly in vapor exposures collected over two weeks.[34] These arrays are capable of identifying a wide range of complex vapor mixtures and volatile organic compounds with classification accuracies upwards of 90%.[34,38] The broad range of chemicals detected with this system, combined with its high classification rates and ability to transfer training information between arrays, highlights the capability of this platform to monitor CWAs in real time.

Processing a complex signal obtained from a cross-reactive sensor often takes computational time and may potentially delay the detection and identification of a harmful vapor. Additionally, a background of non-toxic chemicals at high concentrations could mask harmful vapors making them difficult for the system to detect. In order to avoid these potential problems, it is useful to incorporate sensors that respond specifically and rapidly to harmful vapors into an artificial nose system. To address these issues, chemical switches that exhibit a turn-on fluorescence response have been incorporated into the fiber-optic array platform.[39] Turn-on sensors are more desirable than turn-off sensors due to their low initial background and minimal potential for photobleaching. Bencic-Nagale and co-workers exploited the reaction between non-fluorescent fluorescinamine and acyl or phosphoryl halide nerve agents as the basis for a microbead chemical switch.[39] The reaction produces a fluorescent product, fluorescein phosphoramide (FLPA), and HCl resulting in a 50-fold increase in fluorescence intensity over baseline levels. Acidic conditions were shown to decrease the fluorescence of FLPA, however this decrease was mitigated by coating the microbead sensors with polyvinyl pyridine (PVP), a proton scavenger that maintains basic conditions. These sensors were shown to be specific for diethyl chlorophosphate (DCP), a surrogate for sarin and soman, over other compounds that lack the reactive acyl or phosphoryl halide moiety.

5.5.2 Fiber-optic Biosensors

5.5.2.1 *Acetylcholinesterase Inhibition*

Organophosphorus (OP) compounds are toxic to humans because they inhibit the enzyme acetylcholinesterase (AChE). AChE catalyzes the breakdown of acetylcholine, a neurotransmitter, into choline and acetic acid. A buildup of acetylcholine in the synapses of nerve cells disrupts the normal flow of impulses leading to paralysis and eventually to death. The enzyme is commercially available and has been utilized as the active component of several types of biosensors. Because a by-product of AChE reaction is an acid, pH has commonly been used as a proxy to measure the activity of the enzyme. Most sensors compare the rate of change of H^+ production by the enzyme before and after exposure to an inhibitor. Rogers *et al.* covalently linked a fluorescein isothiocyanate (FITC)-enzyme conjugate to the surface of a quartz fiber in a flow cell and monitored the complex *via* evanescent excitation.[40] Periodic stop-flow quenched the fluorescence intensity of the dye molecule due to the enzyme catalyzed decrease in pH. After an inhibitor passed through the flow cell, the rate of fluorescence quenching markedly decreased or, with more potent inhibitors, did not occur. Detection limits of 5 ppb for the insecticide parathion were reported with a response time as fast as one minute. Höbel and Polster adapted this setup for use on the end of a bifurcated fiber by immobilizing the enzyme along with an FITC-dextran conjugate in a thin polyacrylamide membrane that was draped over the distal end of the fiber.[41] In this sensing configuration, two separate single-core optical fibers are bundled. One fiber functions as an input conduit, delivering excitation light to the sensing region. The second fiber collects and guides the signal back to the detector. A major drawback of this membrane technique was a significant increase in the response time of the sensor, up to 1 hour, making it impractical for real-time monitoring. A more recent approach used an entrapped enzyme, along with FITC-dextran, in a sol-gel matrix that was deposited on the end of a single optical fiber.[42]

Colorimetric measurements have also been employed to measure the change in pH associated with AChE inhibition. Andres and Narayanaswamy immobilized AChE and thymol blue on the surface of separate batches of glass beads that were pooled and placed into a microwell etched into the end of a bifurcated optical fiber.[43] A decrease in the local pH generated a decrease in absorption at 600 nm, corresponding to the non-ionized form of the dye, which was monitored using reflectance measurements.

A deviation from all of the previous pH-based methods measured the absorbance of a colored acetylcholine surrogate.[44] In this system, the synthetic substrate was passed through an enzyme-packed column and the signal of the blue product was measured at 580 nm in a flow cell connected at either end by an optical fiber. Inhibition of the enzyme resulted in a decrease in the intensity of the absorption band. This study also used an LED as the light source, offering the possibility for miniaturization and portability.

5.5.2.2 *Organophosphorus Hydrolase*

Organophosphorus Hydrolase (OPH) is a phosphotriesterase that has been identified in several mammals, insects, bacteria and fungi. It was first purified by Dumas *et al.* in 1989 from the bacterium *Pseudomonas diminuta*.[45] This enzyme catalyses the hydrolysis of several organophosphorus insecticides and chemical warfare agents to less toxic byproducts and acid.[46] A number of studies have investigated the possibility of using this enzyme in the remediation of chemical warfare stockpiles and contaminated areas. Because of its activity to a wide range of OPs, OPH has been incorporated into several types of pesticide biosensors, the majority of which have utilized electrochemical detection.[47]

The gene for the enzyme is encoded on a plasmid, which has led to its expression in a variety of bacterial vectors. The surface-expressed protein on these cells has been used as the active element in several fiber-optic biosensors. Many of these schemes are colorimetric and detect a colored product of enzyme hydrolysis. A recent study utilized *Flavobacterium* cells suspended on glass filter paper that was attached to the end of a bifurcated fiber-optic.[48] The authors were able to detect methyl parathion at a concentration of 0.3 μM by observing the absorbance of its hydrolysis product p-nitrophenol. A separate method employed immobilized *E. coli* cells on an agarose disc that was covered with a porous nylon membrane and fastened to the end of a bifurcated optical fiber.[49] This absorbance sensor, when placed in a flow cell, was able to detect three OPs in a high background of other non-OP pesticides. One drawback of using whole cells suspended in several layers of supporting material was the length of time (30 min) required to obtain a signal. In a effort to decrease the response time, these authors used the same instrumental setup with only the purified enzyme attached to a porous nylon membrane.[50] Streamlining the sensing chemistry to include only the essential components reduced the time required to obtain a positive response to 2 minutes.

The major limiting factor of all the OPH methods discussed thus far has been their dependence on an optically detectable enzyme product. This limitation narrows the scope of potential analytes and, more importantly, poses a challenge to the detection of CWAs where the enzyme hydrolysis product does not absorb in the visible spectrum. One way to expand the range of detectable analytes of this enzymatic approach is to employ an OPH inhibitor that changes its spectral characteristics when it is displaced from the enzyme due to the presence of analyte. Rather than depending on an optically detectable product, the system is tuned to detect inhibitor release due to the higher affinity of the substrate for the enzyme. White and colleagues utilized a weakly bound copper porphyrin inhibitor that changed absorbance characteristics when it was released from the enzyme active site upon the addition of substrate.[51] These spectral differences were detected on a surface functionalized planar waveguide connected, *via* single-core optical fibers, to a visible spectrometer. Detection limits were as low as 7 parts per trillion for paraoxon.

None of the enzyme inhibition-based methods listed above can distinguish between the different organophosphorus nerve agents and almost all are susceptible to false positives induced by non-organophosphorous compounds such as heavy metals or extremes of temperature and pH. The longevity of these sensors is limited to the effective lifetime of the enzyme which, in some cases, is several months. The inhibition-based sensors in general need to be replaced following a positive result, but the sensor life has been extended by regenerating the enzyme with 2-pyridine-2-aldoxime methoiodide (2-PAM). In spite of these advancements, this particular group of sensors has not found widespread use as continuous monitoring devices.

5.6   Conclusions

The methods discussed in this chapter highlight the capability of optical-fiber based approaches to address problems associated with CWA and BWA detection and monitoring. No single platform has emerged as the dominant analytical method; therefore it is necessary to improve upon these existing technologies to enable further detection capability. Recent studies have demonstrated detection limits for microorganisms as low as five cells by using a fiber-optic microarray to detect high copy number ribosomal RNA molecules.[52] This approach eliminates the need for a separate amplification step, thereby decreasing the time necessary to perform a complete confirmatory analysis to 45 minutes. Rissin *et al.* have demonstrated that a microwell array etched into an imaging fiber-optic bundle is capable of detecting analyte concentrations as low as 2.6 amol with enzyme based signal amplification.[53]

These advancements illustrate the ability of optical-fiber based methods to address the response times and low detection limits needed for BWA/CWA detection. New developments in the field of microfluidics should produce more fully integrated systems including sample pretreatment and reagent delivery.

The depth of information content possible with optical-fiber-based methods, combined with their scalability and cost-effectiveness, will make these platforms viable alternatives for BWA and CWA detection well into the future.

Acknowledgements

The authors wish to thank Kyle Bake, Timothy Blicharz, Ryan Hayman and Ragnhild Whitaker for their help in the preparation of this manuscript.

Acronyms

CWA	Chemical Warfare Agent
BWA	Biological Warfare Agent
TIR	Total Internal Reflection
GRIN	Graded Refractive Index Fiber
LED	Light Emitting Diode
CCD	Charge Coupled Device
CMOS	Complementary Metal-Oxide Semiconductor
RAPTOR	Rapid Automatic and Portable Fluorometer Assay System
MANTIS	Man-Portable Total Identification System
UAV	Unmanned Aerial Vehicle
PCR	Polymerase Chain Reaction
MS	Mass Spectroscopy
MIP	Molecularly Imprinted Polymer
SAW	Surface Acoustic Wave
MOSFET	Metal-oxide Semiconductor Field Effect Transistor
QCM	Quartz Crystal Microbalance
DMMP	Dimethyl Methylphosphonate
FLPA	Fluorescein phosphoramide
PVP	Polyvinyl pyridine
DCP	Diethyl chlorophosphonate
OP	Organophosphorous
AChE	Acetylcholinesterase
FITC	Fluorescein isothiocyanate
OPH	Organophosphorous Hydrolase
2-PAM	2-pyridine-2-aldoxime methoiodide

References

1. P. A. Demirev, A. B. Feldman and J. S. Lin, *Johns Hopkins APL Technical Digest*, 2005, **26**(4), 321–333.
2. A. W. Snyder and J. D. Love, *Optical Waveguide Theory*, Chapman and Hall Ltd, London, 1984.

3. R. A. Ogert, J. E. Brown, B. R. Singh, L. C. Shriver-Lake and F. S. Ligler, *Anal. Biochem.*, 1992, **205**, 306–312.
4. L. K. Cao, G. P. Anderson, F. S. Ligler and J. Ezzell, *J. Clin. Microbiol.*, 1995, **33**(2), 336–341.
5. G. P. Anderson, K. D. King, L. K. Cao, M. Jacoby and J. Ezzell, *Clin. Diagn. Lab. Immunol.*, 1998, **5**(5), 609–612.
6. U. Narang, G. P. Anderson, F. S. Ligler and J. Burans, *Biosens. Bioelectron.*, 1997, **12**(9–10), 937–945.
7. T. B. Tims and D. V. Lim, *J. Microbiol. Meth.*, 2004, **59**, 127–130.
8. K. A. Donaldson, M. F. Kramer and D. V. Lim, *Biosens. Bioelectron.*, 2004, **20**, 322–327.
9. I. B. Bakalcheva, F. S. Ligler, C. H. Patterson and L. C. Shriver-Lake, *Anal. Chim. Acta*, 1999, **399**, 13–20.
10. L. C. Shriver-Lake, B. L. Donner and F. S. Ligler, *Environ. Sci. Tech.*, 1997, **31**, 837–841.
11. F. S. Ligler, G. P. Anderson, P. T. Davidson, R. J. Foch, J. T. Ives, K. D. King, G. Page, D. A. Stenger and J. P. Whelan, *Environ. Sci. Tech.*, 1998, **32**, 2461–2466.
12. K. D. King, G. P. Anderson, K. E. Bullock, M. J. Regina, E. W. Saaski and F. S. Ligler, *Biosens. Bioelectron.*, 1999, **14**, 163–179.
13. C. C. Jung, E. W. Saaski, D. A. McCrae, B. M. Lingerfelt and G. P. Anderson, *IEEE Sens. J.*, 2003, **3**(4), 352–360.
14. G. P. Anderson and N. L. Nerurkar, *J. of Immunol. Meth.*, 2002, **271**, 17–24.
15. G. P. Anderson, K. D. King, K. L. Gaffney and L. H. Johnson, *Biosens. Bioelectron.*, 2000, **14**, 771–777.
16. K. D. King, J. M. Vanniere, J. L. LeBlanc, K. E. Bullock and G. P. Anderson, *Environ. Sci. Tech.*, 2000, **34**, 2845–2850.
17. T. B. Tims and D. V. Lim, *J. of Microbiol. Meth.*, 2003, **55**, 141–147.
18. J. M. Simpson and D. V. Lim, *Biosens. Bioelectron.*, 2005, **21**, 881–887.
19. L. Song, S. Ahn and D. R. Walt, *Emerg. Infect. Dis.*, 2005, **11**(10), 1629–1632.
20. L. Song, S. Ahn and D. R. Walt, *Anal. Chem.*, 2006, **78**, 1023–1033.
21. J. A. Ferguson, F. J. Steemers and D. R. Walt, *Anal. Chem.*, 2000, **78**, 5618–5624.
22. H. Chuang, P. Macuch and M. B. Tabacco, *Anal. Chem.*, 2001, **73**, 462–466.
23. A.-C. Chang, J. B. Gillespie and M. B. Tabacco, *Anal. Chem.*, 2001, **73**, 467–470.
24. L. C. Taylor, M. B. Tabacco and J. B. Gillespie, *Anal. Chim. Acta*, 2001, **435**, 239–246.
25. J. Ji, A. Schanzle and M. B. Tabacco, *Anal. Chem.*, 2004, **76**, 1411–1418.
26. L. Bansal and M. El-Sherif, *IEEE Sens. J.*, 2005, **5**(4), 648–655.
27. A. L. Jenkins, O. M. Uy and G. M. Murry, *Anal. Chem.*, 1999, **71**(2), 373–378.
28. A. L. Jenkins, R. Yin and J. L. Jensen, *The Analyst*, 2001, **126**, 798–802.

29. A. L. Jenkins and S. Y. Bae, *Anal. Chim. Acta*, 2005, **542**, 32–37.
30. S. M. Barnard and D. R. Walt, *Environ. Sci. Tech.*, 1991, **25**, 1301–1304.
31. T. A. Dickinson, J. White, J. S. Kauer and D. R. Walt, *Nature*, 1996, **382**, 697–700.
32. J. White, J. S. Kauer, T. A. Dickinson and D. R. Walt, *Anal. Chem.*, 1996, **68**, 2191–2202.
33. T. A. Dickinson, K. L. Michael, J. S. Kauer and D. R. Walt, *Anal. Chem.*, 1999, **71**, 2192–2198.
34. S. Bencic-Nagale and D. R. Walt, *Anal. Chem.*, 2005, **77**, 6155–6162.
35. S. E. Stitzel, L. J. Cowen, K. J. Albert and D. R. Walt, *Anal. Chem.*, 2001, **73**, 5266–5271.
36. K. J. Albert and D. R. Walt, *Anal. Chem.*, 2000, **72**, 1947–1955.
37. K. J. Albert, M. L. Myrick, S. B. Brown, D. L. James, F. P. Milanovich and D. R. Walt, *Environ. Sci. Tech.*, 2001, **35**, 3193–3200.
38. K. J. Albert, D. R. Walt, D. S. Gill and T. C. Pearce, *Anal. Chem.*, 2001, **73**, 2501–2508.
39. S. Bencic-Nagale, T. Sternfeld and D. R. Walt, *J. Am. Chem. Soc.*, 2006, **128**, 5041–5048.
40. K. R. Rogers, C. J. Cao, J. J. Valdes, A. T. Elddfrawi and M. E. Eldfrawi, *Fund. Appl. Toxicol.*, 1991, **16**, 810–820.
41. W. Hobel and J. Polster, *Fresen. J. Anal. Chem.*, 1992, **343**, 101–102.
42. H.-C. Tsai and R.-A. Doong, *Water Sci. Tech.*, 2000, **42**(7–8), 283–290.
43. R. T. Andres and R. Narayanaswamy, *Talanta*, 1997, **44**, 1335–1352.
44. W. Trettnak, F. Reininger, E. Zinterl and O. S. Wolfbeis, *Sens. and Act. B*, 1993, **11**, 87–93.
45. D. P. Dumas, S. R. Caldwell, J. R. Wild and F. M. Raushel, *J. Biol. Chem.*, 1989, **264**(33), 19659–19665.
46. D. P. Dumas, H. D. Durst, W. G. Landis, R. M. Raushel and J. R. Wild, *Arch. Biochem. Biophys.*, 1990, **277**(1), 155–159.
47. J. Wang, R. Krause, K. Block, M. Musameh, A. Mulchandani, P. Mulchandani and M. J. Schöning, *Anal. Chim. Acta*, 2002, **469**, 197–203.
48. J. Kumar, S. K. Jah and S. F. D'Souza, *Biosens. Bioelectron.*, 2006, **21**, 2100–2105.
49. A. Mulchandani, I. Kaneva and W. Chen, *Anal. Chem.*, 1998, **70**, 5042–5046.
50. A. Mulchandani, S. Pan and W. Chen, *Biotechnol. Progr.*, 1999, **15**, 130–134.
51. B. White and H. J. Harmon, *Biosens. & Bioelectron.*, 2005, **20**, 1977–1983.
52. S. Ahn, D. M. Kulis, D. L. Erdner, D. M. Anderson and D. R. Walt, *Appl. Environ. Microbiol.*, 2006, **72**(9), 5742–5749.
53. D. M. Rissin and D. R. Walt, *J. Am. Chem. Soc.*, 2005, **128**, 6286–6287.
54. J. R. Epstein and D. R. Walt, *Chem. Soc. Rev.*, 2003, **32**, 203–214.
55. J. Hecht, *Understanding Fiber Optics*, 4th Ed. Prentice Hall, Upper Saddle River, New Jersey, 2002.

CHAPTER 6

Application of DNA Microarray Technologies for Microbial Analysis

AVRAHAM RASOOLY[a] AND KEITH E. HEROLD[b]

[a] FDA Center for Devices and Radiological Health and NIH-National Cancer Institute, 6130 Executive Blvd. EPN, Rockville, MD 20852, USA;
[b] Department of Bioengineering, University of Maryland, College Park, MD 20742, USA

6.1 Introduction

Since Pasteur and Koch in the nineteenth century, microbial organisms have been identified mainly by microbiological culturing. More recently, additional characterization techniques such as immunological methods and DNA analysis have become more commonly used for microbial identification and typing. Culture based methods are simple to use, relatively inexpensive and sensitive but these time-tested and reliable methods are slow. Typical culture-based assays may require more than a day for pre-enrichment, enrichment and post-enrichment to recover microorganisms. Furthermore, they provide only limited information on the organism and are not very effective for the analysis of multiple microorganisms in environmental samples. For the analysis of microbial pathogens in clinical, food or environmental samples there is often a need to detect small numbers of microorganisms in a large sample volume that may contain interfering substances or multiple organisms. Most other current detection assays (immunological or PCR analysis) are target specific, meaning that the methods can only detect a specific target of interest such as a small group of related organisms or a small number of similar antigens or genes.

Nano and Microsensors for Chemical and Biological Terrorism Surveillance
Edited by Jeffrey B.-H. Tok
© Royal Society of Chemistry, 2008
Published by the Royal Society of Chemistry, www.rsc.org

Though the ability to include a large number of assay results in a single experiment (or test), microarray techniques allow the user to create a broad screening assay not possible with current technologies.

Microarrays, developed in the last ten years, are spatially ordered arrays of recognition ligands (such as oligonucleotide, cDNA, protein, peptide, antibody, carbohydrate, tissue or aptamers) immobilized (chemically bonded) in discrete locations on a solid matrix. The technology is a high-throughput methodology capable of molecular identification and characterization of multiple DNA sequences or proteins in a single array assay. Microarray platforms enable hundreds or thousands of parallel identification assays, each specific to a small section of a genome or a specific antigen. For microbial analysis, such assays provide the ability to obtain multiple detailed genomic or proteomic information regarding the pathogen, including identification of virulence factors or antibiotic resistance, and the capability to analyze multiple organisms simultaneously.

Microarray technology can be traced back to enzyme-linked immunosorbent assays (ELISA) which were developed for specific protein immunological detection.[1] ELISA uses antibodies as ligands to identify target proteins. For DNA analysis, the early origins of microarray technology can be traced to a family of methods for detection of nucleic acid hybridization. Nucleic acid hybridization on a nitrocellulose filter was first developed in 1965 for DNA-RNA hybridization.[2] This was followed by *in situ* hybridization,[3] Southern blot,[4] which is used to detect specific DNA sequences, and Northern blot analysis.[5] The microarray format is based on dot blotting techniques for specific detection of nucleic acid sequences.[6] These early "array" technologies were often used for identification of a small number of targets (*e.g.* in 96 well plates). However, current microarray technology expands the basic approach by reducing the size of the array elements and increasing the number of elements. Ekins *et al.*[7–9] working on "multi-analyte" immunoassays introduced the concept of "microspots" distributed on an inert solid support detected by fluorescence labeling. The technology was originally developed and is mainly used for analysis of gene expression. Beyond that application, microarrays have a significantly wider potential in that they allow rapid microbial diagnostics in a format adaptable to clinical, field and laboratory use. The focus of this paper is on the technology and applications of DNA microarrays for microbial diagnostics.

6.2 DNA Microarray Technology

The original DNA microarray technology was based on immobilization of short sections of nucleic acid (oligonucleotides) on a solid matrix (spatially ordered arrays or chip technology). More recently bead array systems were developed where the capture ligands are immobilized on beads.[10–16] Table 6.1 summarizes some of the characteristics of several common microarray platforms. This manuscript is focused on microarrays on two-dimensional surfaces (*i.e* chip technology).

Table 6.1 Common manufacturing approaches for microarray production.

Method of manufacturing	Probe length, nt	Maximum array spots	Number of slides	Ligand	Development cost	Cost per chip
Beads technology	Any	Less than 100 per well	Use 96 well plate	Oligonucleotide, cDNA, PCR products, protein, peptide, antibody, carbohydrate or tissue	High	High
Spotted arrays	Any	100,000 (Practically several thousands)	Small	Oligonucleotide, cDNA, PCR products, protein, peptide, antibody, carbohydrate or tissue	High	Low
Photolithography	20–25	500,000	Large	DNA	Very high	High
Ink-jet synthesis	60–80	44,000	Medium	DNA	Medium	Medium
Light-directed synthesis	Up to 70	786,000		DNA	Medium	Medium
Light-directed synthesis in nano-chambers	Up to 100	7,963	Large	DNA	Medium	Medium

6.2.1 Spatially Ordered Microarrays

Spatially ordered arrays (*e.g.* DNA chips) are arrays with ligands immobilized (chemically bonded) in discrete locations on a solid matrix. Once the target is bound to one or more of the ligands on the microarray surface, the target's molecular label enables target detection and the binding position on the array enables identification of the sequence. Major steps in microarray analysis are shown in Figure 6.1. All array technologies share several main features: multi-target analysis, specific binding or hybridization of the target and labeling of the target molecules. The two main approaches for DNA microarray production are *in-situ* synthesis of single-strand DNA ligands and robotic spotting of single-strand DNA ligands. In general microarrays can be divided into two classes based on spot density. Low- to medium-density arrays (< 500 spots per array or per cm^2) are mainly fabricated by contact or non-contact printers, providing spot features in the 100–200 µm range. High density microarrays (> 500 spots per array or per cm^2) have spots fabricated by non-contact technologies. For many microbial diagnostic applications, involving an assay for tens or hundreds of specific characteristic sequences, low density arrays are sufficient.

6.2.2 Bead Array Technology

In addition to spatially ordered arrays, microarray bead technology is another high-throughput array methodology. Bead technology is based on internally labeled identifiable polystyrene beads, each with different ratios of two spectrally distinct fluorophores, creating a unique signature for each bead. Each fluorophore can have one of 10 possible levels of fluorescent intensity, and the

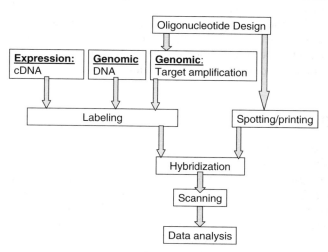

Figure 6.1 Development and application of microarrays for microbial analysis. The main steps for development and application of microarrays for microbial analysis are: oligonucleotide design for spotting and for target amplification (if PCR is used); array spotting and target labeling (cDNA, DNA or amplicons); hybridization scanning and data analysis.

permutations of the two fluorophores on each bead create 100 spectrally identifiable beads. Bead technology has been used for various analysis applications.[16-21] In bead technology, the ligand (such as oligonucleotide or antibody)
is bound to the surface of the beads enabling capture of the target. The target
(*e.g.* DNA, proteins, enzyme substrates, receptors, antibodies) is fluorescently
labeled with another dye (phycoerythrin) which binds to the beads *via* the ligand,
so that each of the ~100 uniquely identifiable beads can bind to a different
target enabling multiplexing. The reader is a flow cytometer with two lasers. A
red diode "classification" laser (635 nm) excites and identifies each of the unique
beads and a second green "reporter" laser (532 nm) detects the presence of the
target on the bead. The system can be applied to a 96-well plate format, where in
each well 100 analytes can be detected with a theoretical throughput of 9600
assay points. This capability is important when a large number of samples
have to be analyzed for many markers. Although spatially ordered arrays can be
divided to allow the analysis of multiple samples, the bead array combined with
the 96-well format provides a very effective multisample analysis.

However, the bead system lacks the high-throughput capabilities of microarrays because each target requires the preparation of specific beads making it
less practical for large numbers of analytes. In addition, microarrays are
manufactured at very low cost per spot and the detector is normally less
expensive than a flow cytometer (low-density microarrays can be analyzed by
inexpensive flatbed scanners).

The bead system was used for several microbial analysis applications
including 16S and 23S gene analysis of four fecal indicating bacteria in river
samples, marine recreational water and beach sand,[22] identification of ascomycetous yeasts from clinical specimens using as a target the species-specific
sequences in domains 1 and 2 (D1/D2) of the large-subunit (LSU) rRNA
gene,[23] and for the identification of *Candida* species and other clinically
important yeast species using a similar approach.[24] Although this manuscript
focuses on chip technology it is noted that many aspects (*e.g.* probe design) are
similar for the two platforms (beads and chips).

6.3   Microarray Design

The design of a microarray and the type of probes used depends on the proposed application of the microarray. The main types of microarrays for
microbial diagnostics are genomic microarrays, which contain sequences
representing the whole microorganism (*e.g.* for gene expression analysis), and
targeted microarrays, for analysis of a limited number of specific DNA
sequences (*e.g.* virulence factors, antibiotic resistance determinants, ribosomal
genes or sequences for mutation analysis). Microarray design includes selecting
target sequences, target functionalization with a label, DNA probe design
(including immobilization method) and spot layout of the array. The focus of
this article is on microbial diagnostics and, thus, probe design for genomic
microarrays for gene expression analysis is not discussed.

6.3.1 DNA Probe Design

A critical element in microarray design is the length of the probes that are immobilized to form the capture array. Microarray probe lengths vary from very long (up to 1000 nt), for cDNA sequences or PCR derived probes, to short oligonucleotide probes (20–70 nt) produced either *in situ* or in bulk oligonucleotide synthesis. The choice of probe length is the primary determinant of the hybridization affinity through its effect on the double-strand melting temperature. In addition to the probe length, the melting temperature also depends on the particular sequence (strong correlation to GC content), secondary structure of the sequence and the experimental conditions such as salt concentration, DNA concentration and the concentration of denaturing agents such as formamide.

Several melting-temperature calculation methods are available and the most accurate are based on the nearest-neighbor thermodynamic model.[25–27] Many computer programs have been developed for probe design including a program enabling overlapping (tiling) probe design.[28] More simplified models for melting temperature determination are based on GC content. These simplified models have substantial limitations compared to the nearest-neighbor model. In addition to melting temperature, another important consideration for oligoprobe design is the potential to form secondary structures such as hairpins, where a single-stranded molecule loops back and anneals with itself. Hairpins need to be avoided in oligoprobe design to maximize hybridization of the target. Sequence analysis of potential probes can easily identify hairpin potential.[28]

Long oligoprobes. Longer oligoprobes (40–70 nt) enable hybridization at higher temperature, reduce background interference and enable hybridization with double-stranded DNA target sequences.[29] Long oligoprobes are used in transcriptomics where quantitative detection of total mRNA is required. The specificity of these microarrays for this application is significantly higher[29] than that obtainable using short oligos, because the hybridization can occur at higher temperature, thus reducing non-specific hybridization in the complex target mixture environment. In an experiment with 50, 60 and 70mer probes, signal intensity increased as the length of the oligonucleotide probe increased, and the 70mer oligonucleotide probes produced signal intensities similar to the intensities obtained with PCR probes.[30] Long oligoprobes are not suitable for discrimination between nearly identical sequences, such as SNP analysis, because the length provides duplex stability even when a single mismatch is present. Other disadvantages of long probes include low efficiency of long oligoprobe synthesis, and high cost of production.

Short oligoprobes. There are three main advantages of short oligoprobe sequences (20–40 nt in length). The first advantage is an enhanced ability to discriminate single nucleotide mismatches between the target ssDNA and the oligoprobe, which allows detection of minor genetic variants in target genes. The second advantage is that shorter oligoprobes allow independent detection of multiple species-specific regions within a single gene enabling redundant coverage of the target sequence with more (but shorter) oligoprobes. The third advantage is that short oligonucleotides are more efficient to synthesize, a factor which may reduce the cost of microarray production.

Short oligoprobes enable discrimination between a perfect match and a single mismatch. With short oligoprobes, even one mismatched base will reduce the melting temperature and consequently reduce hybridization (reduce the signal intensity at a spot). In addition to the factors that influence the stability of perfectly matched sequences, mismatched sequence stability depends on the position of the mismatch within the probe (duplex stability is most sensitive to mismatches in central positions) and the type of mismatch (mismatches involving G are relatively stable).

One approach to enhance discrimination of sequences, even in the presence of some mismatch (such as genes from related organisms that are largely identical but which show some genetic divergence), is tiling, the use of short overlapping oligonucleotide probes[31-35] that provide high specificity through overlapping (redundant) probes.

The main limitations of short oligoprobes are the low melting temperature which reduces specificity, especially in hybridization to double-stranded DNA, and the higher probability of random matches to non-target sequences in the genome. The use of single-stranded DNA targets and redundant probes reduces the probability of misidentification.

6.3.2   Probe Selection

Design of DNA microarrays relies on using bioinformatics tools including genomics databases, DNA homology search tools, secondary structure analysis, and calculation of melting temperature. The software tools commonly used for cross-homology testing of probes against a reference database include BLAST (Basic Local Alignment Search Tool)[36] which enables rapid sequence comparison, and finds alignments that optimize a measure of local similarity. Prediction of secondary structures can be based on a thermodynamic approach, related to the melting temperature determination methods discussed above.

Genomic microarray probe selection.   For microbial genome analysis, computer programs such as Array Designer (Premier Biosoft Intl) (http://www. premierbiosoft.com/dnamicroarray/index.html), ArrayOligoSel[37] and Array-OligoSelector[38] are available. Some of the capabilities of these programs include analyzing the results of a BLAST search performed against a genomic database (*e.g.* NCBI or custom database), identification of significant homologies and repeat regions which are automatically avoided, analysis of the whole organism genome detecting every gene, discovery of differentially and alternatively spliced transcripts, SNP analysis/detection, DNA sequence variation in individuals or populations and comparative genome hybridization (CGH). Such programs are capable of designing thousands of highly specific PCR primers and microarray probes for these applications. Ideally, all of the probes on an array should have similar melting temperatures to avoid temperature-induced variations in hybridization but some of the *in-situ* array manufacturing methods require uniform probe lengths, complicating such temperature-matched design.

A limitation of some of these algorithms is the excessive number of unprocessed BLAST results that complicates final selection of the most specific probes. In addition, some algorithms do not take into account the impact of mismatch position within the probe. Moreover, for bacterial genomes several factors complicate probe design including low GC content[39] and frequent conserved repeats, sometimes leading to erroneous target identification by cross-hybridization.[40]

To overcome these limitations an algorithm for the design of whole-genome microarrays was developed[41] incorporating filtering of oligonucleotide probes libraries sharing homogeneous thermodynamic properties, and annotated probes recognizing highly conserved targets shared by different genomes. This analysis can be performed within a single genome or between several different strains/organisms.

Targeted microarray probe selection. Unlike genomic microarrays, which contain probes which hybridize along the whole genome, targeted microarrays are designed for analysis of a limited number of specific DNA sequences (*e.g.* individual genes, virulence factors or antibiotic resistance determinants) or for mutation analysis.

The design rationale which underlies these microarrays is that, for some applications, there is no need to analyze all the thousands of genes of the organism but, instead, there is a need to know specific information about the microorganism which may be clinically important, such as antibiotic resistance or virulence factors. Moreover, unlike massive genomic microarrays which are time-consuming and expensive to design and build and thus are available in only limited variations, targeted arrays may be easily configured to provide information regarding specific targeted sequences from several species. Targeted arrays have sufficiently low cost to allow multiple variations on the array design during a single investigation.

In terms of array design, the list of target genes can often be generated from the literature so that there is no need for comprehensive genome analysis programs such as Array Designer (Premier Biosoft Intl) but more focused programs such as OligoDesign[28] are sufficient. The criteria for selecting probes are similar to the probe selection for genomic microarray including (i) probe length, (ii) target Tm, (iii) kcal/mol for hairpins, (iv) kcal/mol for self-dimers and (v) the size for dinucleotide repeats.

Software such as OliCheck[41] is designed to test the quality of potential microarray probes by considering the possibility of cross-hybridization with non-target sequences. An example of such an approach was demonstrated for an *S. aureus* microarray.[41] A set of feature elements designed by the program OliCheck was validated experimentally. Probes of length 40–60mer were used combining optimal thermodynamic properties with high target specificity, suitable for genomic studies of microbial species resulting in final selection of 5427 probes yielding >97%, 93% and 81% of *Staphylococcus aureus* genome coverage in strains N315, Mu50 and COL, respectively.

For microarray analysis of gene families present in environmental samples, a program (ProDesign) was developed for designing probes for detecting many gene families simultaneously and specifically in one or more genomes.[42] Gene family-specific probe sequences are generated based on specific shared sequences, which are found with local pairwise alignment. To detect more gene families, common sequences are re-clustered into new families and probes specific to the new families are generated.

6.4 Major Elements of a DNA-Microarray Production

Although many microarrays are today commercially available, the available designs are limited and the cost is relatively high. Many targeted applications require specialized, low-density microarrays which can be produced in-house using equipment whose cost is within the reach of many laboratories. The main equipment required for in-house microarray production and analysis are the arrayer and the scanner.

6.4.1 Solid Matrix for Probe Immobilization

The use of pre-synthesized oligonucleotides is common for low-density micro-arrays used in many diagnostic tests. Several methods have been described in the literature for covalent attachment of modified oligonucleotides to pre-activated solid supports including glass[43,44] or oxidized silicon.[45] Other systems include gold surfaces,[46] optical fibers[47] and plastics (*e.g.* PMMA).[48–50] Three-dimensional matrices such as nitrocellulose or nylon membranes,[51–54] polyacrylamide gel pads[55] and agarose films were used for immobilization of probes.[52]

Two-dimensional glass surfaces are reported to have hybridization advantages compared to those constructed on three-dimensional microporous membranes where it was suggested that solution phase DNA may have greater access to probes immobilized on a planar surface than to those immobilized within a three-dimensional surface. However, other studies show similar performance for two-dimensional and three-dimensional surfaces.[52]

6.4.2 Immobilization of Probes

6.4.2.1 Surface Chemistries for Attachment of DNA Probes

Several surface chemistry approaches were developed for attachment of DNA probes to a solid substrate. These include amine-, aldehyde-, epoxy-, polylysine-groups and dendrimers (PAMAM). Amine-modified glass will adsorb nucleic acids through non-specific electrostatic binding along the DNA backbone. The positive charge of the primary amines attracts DNA usually followed by inter- and intrastrand crosslinking *via* UV activation to immobilize the DNA. A similar approach is used for immobilization of DNA on microporous nylon.[52]

The resulting layer of bound DNA is immobilized in a geometrically complex manner that is not optimum for access by target DNA.

Binding of DNA perpendicular to the surface, in a highly accessible "lawn" of probe molecules, can be achieved by attachment to one end of the DNA strands. A simple example of such a scheme uses glass slides derivatized with 3-mercaptopropyl silane for attachment of 5-prime sulfide-modified oligonucleotides *via* disulfide bonds.[43] Variations on this same theme employ organofunctional silanes as coupling agents. Another variation attaches an aldehyde group to the glass which then will bind covalently to an amino group added to the DNA.

6.4.2.2 *DNA Modifications for Attachment of DNA Probes*

In several platforms, chemical modification of DNA probes with a terminal amine group is required to enable effective covalent binding to the substrate. An amine group is normally added to the 5′ end of the oligonucleotide. For cDNA probes and PCR products, the amine group can be added to the 5′ end of the PCR primer used for the probe amplification. As mentioned above, an alternative is to functionalize the 5′ end of the DNA probe with a sulfide group. Custom oligonucleotides can be ordered with various 5′ end functionalizations for a very reasonable price (on the order of $10 for 25 nmole of a 20 nt oligo).

6.4.3 Printing of Microarrays

For low-density microarrays with a minimum number of spots, contact printing using pre-synthesized ligands is the economic method of choice. For larger microarrays, several other methods based on *in-situ* synthesis of DNA ligands are available. These methods involve synthesis of oligonucleotides on the surface of the chip in multiple cycles where one base (nucleotide) is added in each cycle. Table 6.1 summarizes some of the characteristics of several common microarray platforms.

6.4.3.1 *In-situ Synthesis Methods*

In-situ synthesis methods are based on phosphoramidite chemistry involving elongation of the oligonucleotide (in the 3′ to 5′ direction) on the solid surface by covalent reaction between the 5′ hydroxyl group of the sugar of the last nucleotide to be attached and the phosphate group of the next phosphoramidite nucleotide to be added. In each cycle of elongation only one base is added. To prevent the addition of more than one base during each cycle, the nucleotide added to the growing oligonucleotide has a protective blocking group on its 5′ end. Each elongation cycle includes a deprotection step where the protective group is converted to a hydroxyl group (either with acid or with light) enabling the next cycle of elongation. For microarray applications there are several different methods for deprotection including photodeprotection using masks, photodeprotection with micromirrors and chemical deprotection with synthesis *via* inkjet technology. In general, these methods are far more complicated and expensive than contact printing and thus are more suitable for large-scale, high-density microarray production. Moreover, *in-situ* synthesis technology is limited to DNA microarray production while contact spotting can be used to print many different kinds of ligands such as cDNA, polymerase chain reaction (PCR) products, protein, peptide, antibody, carbohydrate or tissue making the contact printing approach very versatile.

6.4.3.2 Contact Printing

For small microarray projects (several hundreds of spots and printing dozens of slides) contact printing (Figure 6.2) of pre-synthesized oligonucleotides is practical for many small laboratories. Contact printing arrayers for microarray applications were pioneered by Pat Brown's laboratory,[56] which made the technology freely available. Contact printing (or spotting) uses pre-made DNA (or other ligand) that is transferred from a reservoir (*e.g.* microtiter plate) to the surface of the solid matrix using a pin on a robotic arm (Figure 6.2). Glass microscope slides that have been coated with a binding layer are commonly used as the solid matrix. When the pin is touched on the solid surface, a small droplet is transferred creating a circular "spot". Spotting of oligonucleotides is followed by irreversible bonding of the ligand to the binding layer (details depend on the immobilization chemistry).

The robotic spotting system (*i.e.* the arrayer) (Figure 6.2) is relatively simple and inexpensive (on the order of $40K and up depending on features). The pins

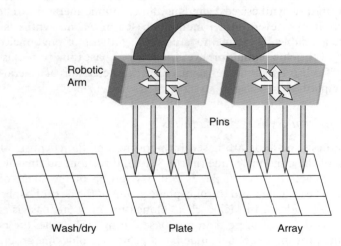

Figure 6.2   Robotic contact spotting technology for DNA microarray production. The pins on the arrayer printing head can be programmed to move in three dimensions. Horizontally, in two dimensions, from the washing/drying station to the oligonucleotide plate and to the array surface. Vertically, dipping in the washing/drying station to prevent cross contamination, contacting the plate to carry the oligonucleotides and to the array to deposit the oligonucleotides.

that carry the DNA from the microtiter plate to positions on the slides are moved in three dimensions (right-left, front-back and up-down) by three computer-controlled stepper motors (motors that can be commanded to move to a particular angular position) connected to linear actuators. Available systems have a typical positional accuracy of $\pm2.5\,\mu\mathrm{m}$. Most print heads can use multiple pins that allow increased printing speed at the cost of programming complexity. The pins on the printing head are moved to washing and drying stations after each spot is printed to minimize cross contamination of the printed oligonucleotides between spots on the array. In many arrayers, humidity is controlled in the printing chamber to prevent drying of the solution during the process. The primary advantage of this type of microarray printing is the ability to create a custom microarray at reasonable cost. The primary disadvantage is that the minimum spot size is larger than the minimum feature size available in the *in-situ* synthesis technologies. Thus, the spot density is small for contact printing. In addition, for contact printing each of the spots has to be printed individually, so printing of large arrays or many arrays is slow compared to the lithography technologies where bases are added to all the spots simultaneously.

Contact printing process. The DNA probes to be printed are organized in microtiter plates (*e.g.* 384 well plates). Many spotting robots can accommodate several plates, allowing printing of arrays with more than 384 spots; alternatively plates can be changed manually. The printing pins used to transfer liquid from the microtiter plates to the glass surface are the most crucial element, designed to deposit tiny drops (nanoliter volume) of liquid on the array. There are a number of different pin designs (Figure 6.2) which determine the size of the spot, uniformity, maximum spot density and the variability among the spots. Solid pins can hold enough liquid (liquid clings to the surface) for one spot on the array. Pins are also available with a reservoir that holds a larger volume, to allow multiple spots to be printed without refilling the pin for each spot. A common pin design is the "quill" type which has a 25-μm-wide slit at the tapered tip of the pin that draws in DNA solution *via* capillary action. The quill-type pin carries and delivers the aqueous ligand solution by a combination of capillary and gravity forces, similar to the physics of a fountain pen.

6.5 Major Elements of a DNA-Microarray Experiment

Regardless of the source of the microarray and the application type, whether gene expression or genotyping, the four main laboratory steps (Figure 6.1) involved in using a microarray are:

1. Target preparation
2. Hybridization
3. Washing
4. Image acquisition and data analysis.

6.5.1 Target Preparation

The term "target" refers to the sample being probed by the microarray. Microarray targets are often PCR amplicons, genomic DNA, cDNA or total RNA. To be detected, the target incorporates either molecules of a fluorescent dye or other detectable marker, such as biotin, that permits subsequent detection with a secondary label. Due to ease of use, fluorescent labeling technologies are a convenient method for the detection of hybridization on microarrays.

The most commonly used and widely commercially available fluorescent labels include Cy3 (excited by a green laser) and Cy5 (excited by a red laser). These fluorescent dyes are excited by most commercial laser scanners and possess high extinction coefficients (150 000 L/mol-cm for Cy3 and 250 000 L/mol-cm for Cy5), providing high sensitivity and broad dynamic range. For expression experiments measuring differential gene expression, the two labeled samples (each labeled with a different dye) to be compared are mixed and hybridized to an array. The labeling with different dyes allows differential analysis of expression in the two samples by comparison of the relative intensity of the two signals at each spot, leading to the familiar checkerboard microarray picture containing two primary colors plus combinations of the two. For genotyping, such multiple dyes are important for quality control purposes.

Hybridization of the target depends on its characteristics including: the type of ligand (DNA or RNA), type of molecule (single or double stranded) and size (long or short sequence). These characteristics are critical for selecting the best combination of microarray platform and experimental conditions. For example, we found that low concentrations of long (2 kb) double-stranded PCR amplicon do not produce a strong hybridization signal with short oligoprobes (25 nt). Approaches to increase the signal in such a case include using a higher concentration of the target DNA, converting the double-stranded molecules to a single stranded target, target length fragmentation to approximately 100–200 bp, adjusting stringency (by changing both salt (or formamide) concentration and temperature), or using longer oligoprobes.

6.5.2 RNA Target Preparation

There are a number of different ways in which an RNA sample can be prepared and labeled for microarray detection, such as gene expression analysis. The basic methodology for microarray expression analysis is described by Hegde *et al.*[57] An important element for RNA preparation is the removal of DNA *e.g.* using on-column DNase treatment. In eukaryotic systems there are many established protocols for amplification of mRNA, which rely on the poly(A) tails. Amplification of prokaryotic mRNA, which lacks poly(A) tails, is more challenging and has limited the application of microarrays in microbial gene expression analysis.

6.5.2.1 RNA Isolation

The first step is extraction of total RNA from the tissue of interest. The protocol is usually based on total RNA extraction from cultures, using one of many kits for RNA purification. An example of such an RNA preparation for gene expression analysis[41] is based on total RNA extraction from cultures, combining guanidine-isothiocyanate lysis with silica-gel–membrane purification.

Since the majority of RNA in a typical cell is rRNA, it is desirable to separate the mRNA prior to the microarray experiment. Isolation of mRNA from eukaryotic sources has been performed using oligo(dT) selection. Bacteria, however, lack the poly(A) tails found on eukaryotic mRNA making isolation of mRNA from bacteria more complicated. A method for removal of rRNA from microbial total RNA based on capture oligonucleotides that bind to the bacterial 16S and 23S rRNAs was reported.[58]

6.5.2.2 RNA and cDNA Amplification

RNA can be converted to cDNA, using reverse transcriptase that synthesizes a complementary DNA (cDNA) strand from single-stranded RNA, and then amplified using PCR (so-called RT-PCR). Eukaryotic mRNA can be primed with a poly-T primer starting the reverse transcription from the 3' end of the mRNA. For bacterial mRNA, the RT-PCR process can be primed using random primers. However, cDNAs are often biased toward the 3' end of the mRNA due to transcript length effects. To overcome such problems, the probes on the array can be biased toward the 3' region of the mRNA.

To increase sensitivity, RNA can be primed with a synthetic oligonucleotide containing the T7 RNA polymerase promoter sequence 5' to a polythymidylate region.[59] After second-strand cDNA synthesis, T7 RNA polymerase was used to generate amplified antisense RNA (aRNA), which has been shown to retain information on transcript abundance[60] and is especially useful when the amount of RNA available for gene expression profiling is limited. However this approach is limited to eukaryotic systems because it relies on the poly(A) tails.

Another approach for RNA amplification when limited RNA is available is the use of Linear Amplification of Prokaryotic Transcripts (LAPT). This method uses the overhang tailing activity of Moloney murine leukemia virus reverse transcriptase to add the T7 promoter to the 5' end of cDNA during reverse transcription, enabling the unbiased amplification of sense-stranded RNA in the presence of mammalian RNA (at a eukaryotic/prokaryotic RNA ratio of 500 to 1). This technology enables microarray analysis of prokaryotic transcriptomes.[61]

6.5.3 RNA Target Labeling

Almost all detection of cDNA is based on fluorescent labeling; colorimetric detection labels (the products of horseradish peroxidase, alkaline phosphatase, gold-silver staining, *etc.*) have also been used in limited applications and will not be discussed here.

6.5.3.1   Direct Incorporation

For eukaryotic microorganism RNA (*e.g.* fungi and yeast), the most common labeling method is direct incorporation of fluorescently labeled nucleotides by reverse transcriptase. In the presence of labeled cyanine dCTP, the RT-PCR amplification produces cDNA with labeled bases at each cyanine location in the sequence. A limitation of this approach is that the incorporation of cyanine labeled dCTP is normally less efficient than unlabeled dCTP, reducing the yield of the reverse transcriptase reaction. In addition, while Cy5 is incorporated less effectively by reverse transcriptase than Cy3, the extinction coefficient of Cy5 is higher than Cy3, introducing an intensity bias in the double-label experiment.

6.5.3.2   Indirect Labeling

Indirect labeling of reverse transcription products through incorporation of an amino-allyl-modified dCTP during cDNA synthesis, followed by reaction of the resulting cDNA with an active ester of the dye, overcomes some of the limitations of direct labeling. Because amino-allyl-dCTP is more similar to the unmodified base (dCTP) as compared to the Cy-modified dCTP, reverse transcriptase incorporates it normally resulting in higher total yield of the reaction. Moreover, the rate of incorporation of Cy3 and Cy5 into the cDNA using this indirect scheme is similar, thus eliminating the incorporation bias found for direct labeling.

6.5.3.3   Chemical Labeling

Chemical processes can be used to add fluorescent cyanine labels to RNA (or cDNA) directly, thus making the reverse transcription step unnecessary.[62] One problem with labeling RNA directly is the orientation of the labeled probe. This method labels the actual RNA from the sample (the sense strand). Commercial oligonucleotide microarrays normally use probes based on the sense strand, assuming an RT-PCR step that creates anti-sense target molecules. Thus, many commercial microarrays may be incompatible with this labeling method.

A technique for direct chemical labeling of DNA with Cy3/Cy5 bearing alkylation agents was described.[63] An advantage of this method is that the labeling can be performed at any step of the assay before the hybridization, considerably increasing flexibility of a microarray experiment.

6.5.3.4   Nanoparticle Labeling

Nanoparticle labeling methodologies[64] utilize gold nanoparticles derivatized with a poly-T eukaryotic mRNA binding tag to detect targets captured onto oligonucleotide microarrays. Because of their unique properties, nanoparticle labels can be visualized by optical (absorbance, resonance light scattering, colorimetric, surface enhanced Raman spectroscopy) or electrochemical means. A silver development methodology can be utilized to amplify the signal associated with the particles by over five orders of magnitude. Compared to an unamplified Cy3 fluorescence assay, silver amplified gold nanoparticle detection provides a ~1000-fold increase in sensitivity.

6.5.3.5 Post-hybridization Labeling

This approach does not require labeling or amplification of the target and thus minimizes some of the problems with target labeling. After hybridizing the cDNA (or RNA) to an oligonucleotide microarray, the bound molecules are detected in a second hybridization step using oligonucleotide (oligo-dT$_{20}$)-modified gold nanoparticle probes[65] which hybridize to the poly(A) tail of the captured mRNA molecules (method only for eukaryotic mRNA analysis). The method increases sensitivity ~1000 times compared to un-amplified fluorescent-based methodologies.

6.5.4 Considerations for Prokaryotic RNA for Microarray Analysis

As discussed above, in many protocols for RNA isolation, mRNA from eukaryotic sources can be isolated using oligo(dT). However, selection using such an approach will not work for bacteria which lack the poly(A) tails found on eukaryotic mRNA. For bacterial sources, this complicates the labeling and isolation of bacterial mRNA and the generation of cDNA for microarray applications and requires the development of new approaches for bacterial RNA analysis.

6.5.4.1 Isolation of mRNA from Bacteria

An approach to isolate bacterial mRNA is to remove the 16S and 23S rRNA (which is 80% or more of a bacterial RNA) from total RNA of bacterial species. One method is based on using capture oligonucleotides that bind to the bacterial 16S and 23S rRNAs.[58] A similar magnetic capture-hybridization method was developed for purification of bacterial mRNA from total RNA by removing 5S rRNA in addition to the16S and 23S rRNA.[66]

6.5.4.2 Generation of Bacterial cDNA and cDNA Labeling

For generating eukaryotic cDNA the mRNA can be primed with a poly-T primer starting the reverse transcription from the 3' end of the mRNA or by using random primers. For bacterial mRNA, the only option is random primers. Poly-T priming provides significant flexibility and convenience over random priming which may suffer from lower yields due to the lower primer concentrations possible when the entire random set of primers must be provided. Another feature of poly-T priming is that it works well even for small amounts of target. However random priming, unlike Poly-T priming, generates a more uniform representation of all the transcripts and is not biased toward the 3' end. Once the cDNA is generated, any of the above methods for cDNA labeling can be used.

6.5.4.3   Presence of Host RNA Mixed with the Bacterial RNA

Another challenge in working with bacterial RNA is the presence of host RNA mixed with the bacterial RNA in clinical or research samples, as host (*e.g.* human) RNA can compete with bacterial RNA during cDNA synthesis. An approach to avoiding this contamination involves cell specific lysis. For analysis of gene expression of *Escherichia coli* interacting with human brain microvascular endothelial cells,[67] it was possible to eliminate the human RNA by cell-specific lysis of human cells which did not lyse the bacterial cells. The intact bacteria were then separated by centrifugation followed by RNA extraction and microarray analysis.

6.5.5   DNA Target Preparation

Unlike mRNA used for expression analysis, genotyping uses genomic DNA as the target, which is usually a double-stranded molecule that can interfere with hybridization of the target DNA to the microarray probe. Although the whole genome can theoretically be used as the target, this is not common practice. To improve hybridization efficiency, DNA can be either fragmented or converted to a single-stranded DNA (ssDNA) prior to hybridization. To increase sensitivity, microbial DNA is often amplified by PCR or, more recently, by whole genome amplification (WGA).

6.5.5.1   Whole Genome Amplification (WGA) Technologies

When a microbial source is limited (less than 1 μg) some type of amplification must be employed prior to hybridization. If a sample contains viable cells, it can often be cultured to provide amplification. In the absence of viable cells, WGA protocols can be used to obtain enough target DNA for a microarray experiment. There are three WGA amplification approaches that are useful to complement detection and identification of bacterial pathogens. All three WGA methods are based on extension of random primers (6–8mers) annealed at random locations along the genome, resulting in amplification (10–10 000-fold) of the entire genome with varying levels of bias.

- Random hexamers, octamers or other random primers such as degenerate oligonucleotide primed PCR (DOP-PCR)[68] was one of the first methods used for WGA amplification with Taq DNA polymerase.
- Multiple-displacement amplification (MDA)[69–71] uses the highly processive bacteriophage Phi29 DNA polymerase and random exonuclease-resistant primers (to protect them from degradation during the amplification) in an isothermal amplification reaction. In contrast to the Klenow fragment, the strand-displacement synthesis with the Phi29 polymerase generates long DNA products (> 10 kb in length). Another advantage is that Phi29 DNA polymerase has higher proofreading activity resulting in lower misincorporation rates compared to other DNA polymerases.
- OmniPlex converts randomly fragmented genomic DNA into a library of DNA fragments which can be amplified by PCR.[72,73] This approach consists of initial DNA fragmentation followed by adapter ligation to form a genomic library which can be amplified by PCR using primers complementary to the adapter sequence. This technique results in a range of fragment lengths (between 150 and 2000 bp) depending weakly on the amount of sample.

6.5.5.2 Amplicon Fragmentation

To increase the strength of hybridization, especially for long double-stranded DNA molecules hybridized with short oligoprobes, the target DNA must be fragmented before hybridization. Three methods can be used for DNA fragmentation: A) physical shearing (*e.g.* sonic fragmentation), B) chemical shearing and C) partial digestion with a restriction enzyme that cuts relatively frequently within the genome or with a randomly cleaving enzyme such as DNase I.

6.5.5.3 PCR Amplification of Target DNA

For analysis of a small number of target sequences within a bacterial genome, PCR amplification is a practical, sensitive and specific target preparation procedure for the detection of genetic markers of interest.[74–84] PCR usually results in > 100 000-fold amplification of target sequences, even in the presence of a much larger amount of non-specific DNA, enabling detection of a single bacterial cell in complex biological, food or environmental samples.

Amplification products resulting from PCR are usually double-stranded DNA molecules. Several methods were developed for utilization of PCR amplicons for hybridization:

- Double-stranded PCR products can be used directly for hybridization on microarrays[79] although the signal is often poor. Long oligoprobes (> 25 nt) improve hybridization in such an approach.
- Fragmentation of double-stranded target DNA (*e.g.* using DNase) is often necessary for use with microarrays made with short oligoprobes (20–25 nt).

- Single-stranded target DNA is a more efficient target form for hybridization as compared to double-stranded DNA. In particular, single-stranded preparation is often the method of choice for use with short oligoprobes. Single-stranded target can be produced directly by asymmetric PCR,[84] or obtained from PCR amplicons by a primer extension reaction,[83] strand separation using biotinilated primer and streptavidin-coated magnetic beads or *in-vitro* transcription by RNA-polymerase utilizing T7 promoter tag attached to the 5'-end of the primer, which is serving as a recognition site for RNA polymerase.[77] This method is recommended for obtaining very long single-stranded targets (up to 10 kb).

6.5.6   DNA Target Labeling

The focus of this section is on fluorescent labeling since this is the most commonly used labeling method. Colorimetric detection labels such as the products of horseradish peroxidase, alkaline phosphatase and gold-silver staining can also be used but since these methods are less common they will not be discussed here.

The labeling methods already discussed for cDNA targets derived from mRNA are all applicable for labeling targets derived from genomic DNA. These include direct chemical, direct incorporation during PCR, indirect chemical and nanoparticle labeling. In addition to these methods, random priming is used with long targets such as genomic DNA.

Random priming is a common way of labeling genomic DNA. DNA is incubated with random primers and extended using the Klenow fragment of DNA polymerase I in the presence of labeled nucleotides. The strand displacement activity of the Klenow fragment generates a mixture of short, single-stranded, labeled target sequences complementary to both strands. The isothermal strand displacement procedure usually results in 20–150-fold amplification of the starting DNA. The source material can be genomic, plasmid or total DNA. This approach works well when the starting amount of bacterial DNA is greater than 1 μg.

6.5.7   Hybridization

In the hybridization step, the DNA probes on the microarray substrate and the complementary labeled DNA (or RNA) target anneal to form a double-stranded molecule. Following annealing, unbound target is then washed off the array leaving labeled target tethered to the surface at positions indexed to known probe sequences.

Hybridization kinetics depends on the DNA probe surface density and length of the immobilized DNA but is largely independent of immobilization substrate.[52,85] Hybridization kinetics can be described as a function of the immobilized DNA density by fitting the data to the equation:[48]

$$V_0 = \frac{V_{\max}I}{K + I} \qquad (6.1)$$

where V_0 is the initial hybridization rate, V_{max} is the maximum calculated hybridization rate, I is the surface density of immobilized DNA probes and K is the probe surface density where the hybridization rate is half of the maximum.[85] Experimentally, plotting the initial rate of hybridization for different immobilized capture probe densities demonstrates that the hybridization rate is dependent on the DNA surface concentration.

Hybridization conditions determine non-specific binding and background level. The concept of "stringent" hybridization conditions refers to conditions that are largely unfavorable to hybridization, with the idea being to minimize non-specific binding to the probes while capturing only the complementary sequences to the probes. Several parameters affect hybridization stringency including: temperature, salt concentration, formamide concentration, probe size, target concentration, hybridization chamber configuration and time.

The primary variable in a hybridization experiment is temperature with high temperature being a stringent condition. In general, high salt concentration stabilizes duplex formation and thus decreases stringency. Na^+ is commonly used in the hybridization solution at 1 M concentration. Formamide, which is a highly polar solvent, destabilizes duplex formation (increases stringency). Formamide is used in many protocols as it tends to denature DNA and disrupt secondary structure, increasing the probability of hybridization to a complex target. The typical range of formamide concentrations used is between 0 and 50%. To reduce non-specific annealing of the target DNA, carrier DNA may be added to the hybridization solution. Hybridization reactions can take 12 to 24 hours, with shorter times (a few hours) for short probes, short target lengths and high target concentration.

Common hybridization protocols utilize static incubation of the target in the hybridization solution, with the array in a hybridization chamber. Hybridization kinetics is governed by diffusion from the liquid to the surface. The hybridization chamber is typically a small cassette. After loading the sample, the hybridization chamber is sealed and placed in an incubator or a temperature-controlled water bath. The speed of the hybridization reaction can be increased through mixing. Other factors affecting hybridization include the array feature size, which influences nucleic acid surface capture in DNA microarrays.[86]

For high-throughput systems, robotic hybridization stations have been developed to handle multiple chips automatically, including incubation and washing steps. Such automation may reduce the variability of microarray experiments.

6.5.8 Washing

To remove unbound target and excess hybridization solution from the array and to assure the specificity of binding so that only the DNA complementary to each spot will remain bound after hybridization, the microarray is washed after hybridization. Washing is normally done with a series of increasingly stringent solutions designed to reduce cross-hybridization and minimize background. Our protocol involves washing with a low salt-detergent buffer (*e.g.* 0.1-X SSC and 0.1% SDS) or with a high-temperature wash in several washing cycles each with decreasing salt concentration (increasing stringency).

6.5.9 Array Scanning

Once the fluorescent target is hybridized to the microarray and the unbound material is washed away, the labeled target bound to each element on the chip is detected by scanning the surface for fluorescence. In general, the scanners used for array analysis are optical devices which image the array surface and record fluorescent intensity as a function of position in two dimensions. Three types of scanners have been used successfully for microarray detection: confocal scanning devices, CCD cameras and flatbed scanners. All such systems excite the fluorophore and then measure the fluorescent emission intensity and convert it to a digital array image. Each pixel on the scanned image represents a single point of measurement. The minimum detectable element size ranges from 25 to 500 μm in diameter, depending on the detector and optics (with typical spotted arrays measuring 100 μm in diameter). A rule of thumb is that the spatial resolution of a microarray scanner should be less than 1/10 of the diameter of the smallest microarray spot. Thus, 100-μm microarray spot diameters require 10-μm spatial resolution to provide a convincing and repeatable result.

6.5.9.1 Laser Scanners

The most common scanners are based on confocal laser scanning which provides high-resolution imaging by scanning a very narrow depth of focus which limits background artifacts. Typical commercially available confocal imaging systems provide 5–10 μm resolution. The slide image is scanned by moving the slide or the confocal lens (or both). Most systems utilize two lasers, including a green laser (for Cy3; excitation wavelength is 550 nm and emission wavelength is 581 nm) and a second red laser (for Cy5; excitation wavelength is 649 nm and emission wavelength is 670 nm). To scan for both colors, some systems scan the array twice while others collect both signals in a single pass. Light emitted from the fluorescent sample at each spot is filtered, and the light is collected with a photomultiplier tube (PMT) or similar detector, and converted to an electrical signal.

6.5.9.2 CCD Scanners

Fluorescent or colorimetric imaging with a CCD (charge coupled device) camera-based scanner can be done economically by leveraging the recent rapid advances in CCD camera technology for the consumer market. Many CCD-based scanners utilize continuous wavelength light sources (*e.g.* arc lamps), enabling the detection of more dyes and eliminating the need for multiple lasers. Filtering of emission spectra minimizes optical cross-talk between different fluorophores. Unlike laser scanning, which focuses the excitation on a small area of the array, CCD-based imaging typically involves illumination and detection of a large portion of the slide ($\sim 1 \, cm^2$) in each image. This has the advantage of reducing the complexity of the scanning system, as compared to the confocal system, thus simplifying instrument design and reducing cost. The primary limitations of CCD scanners are generally a lower spatial resolution (*i.e.* approximately 20 to 50 μm) and the relatively broad depth of focus which can detect background artifacts.

6.5.9.3 Flatbed Scanner

An even simpler approach for microarray imaging is the use of an off-the-shelf desktop flatbed scanner. Systems based on flatbed scanners often utilize colorimetric detection labels such as the products of horseradish peroxidase, alkaline phosphatase or gold-silver staining. The spatial resolution of such scanners is 5–50 μm. These low-cost scanners can detect a single color only due to lack of filtering and are primarily used for low-density microarrays.

Several examples can be found in the literature using flatbed scanners. A flatbed scanner was used for cDNA detection with cationic gold nanoparticle labels (with diameters of 250 nm). Sensitivity was estimated to be less than 2 pg of DNA molecules captured on the array surface.[87] The approach utilizes non-labeled target molecules hybridizing with complementary probes on the array, followed by incubation in a colloidal gold solution. The hybridization signal results from the precipitation of nanogold particles on the hybridized spots due to the electrostatic attraction of the cationic gold particles and the anionic phosphate groups in the target DNA backbone. A flatbed scanner was used for microarray-based quantitative gene expression analysis.[88] In this study, the target cDNA was labeled with biotin and was detected by streptavidin-conjugated alkaline phosphatase staining.[89]

6.5.10 Data Analysis

The scanner output of a microarray is usually a combination of two monochrome images: one for each of the two wavelengths measured. For two-color differential gene expression studies, these images are combined to create red-green-yellow false color microarray images, with yellow used when both wavelengths are present. The dynamic range of the detector depends on the technology, the system electronics and the software. Typically, a 16-bit TIFF image format is used, which takes values between 0 and 2^{16} (65 536). With a background cutoff of ~ 100 and saturation intensity of $\sim 50\,000$, the microarray system can detect intensities over an approximately 500-fold dynamic range.

Expression microarrays generate a large number of data points for each experiment. There are many sources of variability in expression microarrays, in addition to the differential gene expression which is sought. These include the probe melting temperature, the quality of RNA, printing conditions, labeling and hybridization conditions. As a result, data analysis for microarray differential gene expression experiments requires a sophisticated bioinformatics approach. Detailed aspects of the bioinformatics related to expression microarrays were reviewed recently[90] and are beyond the scope of this manuscript, which is focused on genomic microarrays.

In general, expression microarrays can be analyzed using several approaches to determine the association of genes which can be either "supervised" or "unsupervised". Supervised methods require a pre-existing classification from outside the microarray data set to be analyzed (*e.g.* a subset of data used as a training set, knowledge of gene function or regulation, phenotype, tissue origin or cell type). In unsupervised methods, there is no pre-existing classification and no additional information besides the expression data itself is used. Such methods are geared towards uncovering expression clustering patterns in the data. The main purpose of clustering methods is to group genes based on similarity of expression profiles (*i.e.* genes that are expressed together most frequently). Clustering provides a useful tool for extracting underlying gene expression information. Many clustering procedures have been developed.

Unlike expression arrays where the main question is the association of genes, microarrays for genomic analysis are used to assay the presence of genes or alleles. As a result, the data analysis and bioinformatics is far less complex for genomic arrays. The presence of fewer spots on typical low-density genomic arrays limits the amount of data. Typically, the primary data from the array consists of intensities of a single fluorophore instead of a two-color experiment. In our work, we have used a second color scan to improve quality control but the primary data are still monochrome. Although these features imply less data output than would come from expression array studies, there is still a need for careful documentation and archival of data to allow useful interpretation.

The technology and the technique are relatively complex with a large number of variables that can influence the results. Many discrepancies in microarray assay results have been reported, especially when using different microarray platforms.[91–93] However, for genomic analysis, reproducibility can be obtained through design of the array and the technique with an emphasis on quality control.[77,81,83]

6.6   Applications of Microarray Technologies for Microbial Analysis

The power of microarray technology is the ability for simultaneous analysis of a large number of specimens and a large number of molecular markers. The two most common applications of DNA microarray technology are analysis of gene expression and genotyping (targeted analysis).

For targeted analysis, this capability enables many potential applications including microbial genotyping, microbial identification, DNA resequencing, mutational analysis, microbial community population analysis, microbial ecology analysis, antibiotic resistance determination and virulence factor identification. These applications are relevant to several fields including environmental microbiology, microbial ecology; human medicine, veterinary medicine, food safety, plant biology; water quality control; industrial microbiology, vaccine analysis, medical device contamination, molecular epidemiology, evolution studies and microbial physiology.

These applications have potential use in research, clinical, agricultural, regulatory, public health (for epidemiological investigations), industrial and ecological settings. Microarray gene expression analysis is well described in the scientific literature so it is only summarized here.

6.6.1   Microarray Analysis of Gene Expression

The original and the most widespread application of microarray technology is the study of gene expression, where microarrays are used to analyze differential gene expression and to compare and quantify the relative abundance of mRNA between samples (Figure 6.3).

To obtain the two samples for a differential analysis, cell extracts are processed to yield complete and intact mixtures of mRNA from each cell. The mRNA in each of the two samples is then converted to labeled cDNA by reverse transcriptase, using one of two fluorescent dyes. The two labeled samples are then mixed and hybridized to the microarray. Following hybridization, the array is washed to remove everything except tightly bound target molecules and scanned using excitation for both dyes. Data analysis of the fluorescent signals is performed for each of the microarray spots. The relative intensity of each of the dyes is measured and the relative abundance of each mRNA is calculated. Expression microarrays generate a large quantity of data since there is a large number of spots and each spot has two intensity values.

6.6.2   Microarray Targeted Analysis and Genotyping

Genotyping is an important application of microarrays used to determine whether a specific sequence is present in a genomic sample (Figure 6.4). This can include analysis of an entire genome, a subset of genes in a genome, a set of genes from several genomes or multiple alleles of a single gene such as single nucleotide polymorphisms (SNP). Unlike expression microarrays, which start from mRNA, a genotyping analysis typically starts from genomic DNA which is extracted from a cell, purified, amplified using PCR or whole genome amplification (if necessary), and labeled with a fluorescent dye.

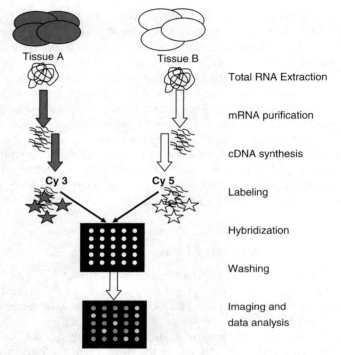

Figure 6.3 Microarray for differential gene expression analysis. Cellular extracts from two tissue samples to be compared are processed to obtain a purified mixture of total RNA, mRNA is purified and is converted to cDNA by reverse transcriptase. The cDNA in each of the two samples is labeled with a fluorescent dye (*e.g.* Cy3 and Cy5) unique to that sample. The resulting labeled samples are then mixed and hybridized to the microarray. Following washing, the array is scanned for both dyes and the fluorescent signals are analyzed, and used to compute the relative intensity of each of the dyes to calculate relative abundance of each mRNA followed by cluster analysis to group genes with common expression pattern.

Unlike expression microarrays, only one dye is necessary for genotyping analysis. However, a second dye can be used for quality control purposes. In several applications[77,83,94] a specific quality control oligonucleotide, labeled with a different fluorescent dye, is mixed with the target DNA. The microarray is printed with each spot containing a mixture of the complementary sequence of the target genes and the quality control oligonucleotide (complementary to the oligo mixed with the sample DNA). When the mixed DNA sample is then hybridized to the microarray followed by washing, scanning and data analysis, the resulting image has a signal at every spot on the microarray for the quality control dye. This quality control feature can be used to verify proper printing and hybridization of the microarray, which is important for clinical and regulatory applications. For genomic analysis the main question is whether or not

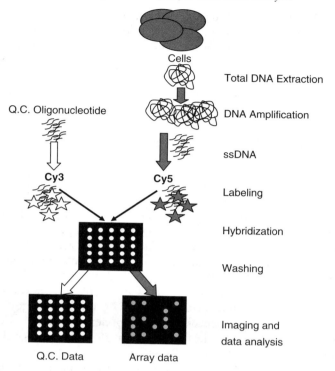

Figure 6.4 Microarray microbial genotyping analysis. Genomic target DNA is extracted from a cell, amplified (if needed), converted to single-strand DNA (if needed). The DNA can be labeled during or after amplification with a fluorescent dye (*e.g.* Cy5). In some applications, for quality control, the labeled target DNA is mixed with a quality control oligonucleotide (complementary to QC oligonucleotide printed in each spot) which is labeled with a different fluorescent dye (*e.g.* Cy3). Target DNA (or the mixed DNA sample) is hybridized to the microarray followed by washing, scanning and data analysis. A quality control scan can be used to verify proper printing and hybridization of the microarray. The resulting image has target and QC signals at every spot on the microarray and can be used to identify the presence of the target DNA in the sample and the quality of the microarray.

a particular gene or allele is present. Several applications of microarrays for genomic analysis and genotyping are described here.

6.6.2.1 DNA Sequences for Microbial Analysis

For microbial microarray genotyping analysis, it is feasible to probe every gene in the genome on a single array. However, more focused arrays are also useful that probe signature sequences from the genome. Several genomic characteristics (signature sequences) have been identified and detected including genes coding for rRNA, toxins, shared conserved indels (insertions/deletions),[95-98] virulence factors and antibiotic resistance potential. One approach to generating signature sequences is multiple alignment between sequenced genomes, which often yields conserved sequences between organisms. The conserved sequences can sometimes be used to design universal PCR primers that span regions that exhibit species-specific variation useful for species identification.

6.6.2.2 Ribosomal DNA Polymorphisms for Bacterial Identification

Ribosomal RNA and corresponding genes (*rrn*) can be used for bacterial identification and evolutionary classification based on several key characteristics[99] including: (i) ribosomal RNAs are ubiquitous to all organisms, and show significant structural and functional conservation, even between divergent species, (ii) ribosomal RNAs are abundant, readily isolated and identified, (iii) structurally they contain both variable and highly conserved regions making them easy to amplify by PCR (using primers designed for the conserved regions) and easy to identify using the variable regions, (iv) the variable regions exhibit conservation within a species and they do not exhibit horizontal gene transfer, and (v) the *rrn* sequences of many species are publicly available. In prokaryotes, the *rrn* loci contain the genes for all three conserved ribosomal RNA sequences (16S, 23S and 5S) separated by highly variable spacer regions.[100] The 16S rDNA is the most common ribosomal sequence used;[101] however, it was suggested that the 16S rRNA gene does not typically allow resolution below the species level.[29,102] An Affymetrix GeneChip with over 30 000 microbial 16S rDNA oligoprobes was used to identify 17 of 19 populations of airborne bacteria, to the level of higher phylogenetic taxa.[103,104] The 23S rDNA was used for microarray analysis of blood pathogens[101] and for bacteria causing infertility and abortions in mares.[105] The variability of 23S rDNA is similar to that of the 16S-23S rDNA spacer region[101] but the 23S rDNA variable region is larger than that of the 16S-23S rDNA spacer region, making it more suitable for bacterial identification. The 16S-23S rDNA spacer region has also been used for microbial classification. Examples of use include identification of *Bacillus anthracis*,[106] targeting the 16S rRNA and 16S-23S rRNA intergenic region, and for Campylobacter identification.[107]

6.6.2.3 Microarray Identification of Bacterial Virulence Factors and Antibiotic Resistance Determinants

Microarrays can be used for identification and characterization of microbial pathogens, including identification of microbial virulence factors and antibiotic resistance using signature sequences and characteristic genes. Identification of microbial virulence factors and antibiotic resistance has clinical importance as well as epidemiology applications. Some of the genes for these characteristics are found on plasmids prone to horizontal gene transfer and such transfer can be diagnosed and observed using an appropriate microarray system.

6.7 Examples of Applications of Microarrays for Microbial Analysis

Several example applications of microarray-based microbial genotyping are described here, including microbial identification, microbial community population, antibiotic resistance, virulence factor and food safety analysis. Such microarray analysis methods have potential use in research, clinical, agricultural, regulatory, public health, industrial and ecological settings.

6.7.1 Microbial Identification and Characterization

The application which pioneered the use of microarray technology for microbial characterization was created at the FDA laboratories with the aim of discrimination among food pathogens including *Escherichia coli* and other pathogenic enteric bacteria harboring various virulence factors.[79] The basic technology used in this and other related work was to use PCR to amplify clinically relevant target sequences (*e.g.* virulence factors) which were then identified by hybridization to microarray.

6.7.1.1 Analysis of Food and Water Borne Microbial Pathogens

The presence of six genes (eaeA, slt-I, slt-II, fliC, rfbE and ipaH) encoding bacterial antigenic determinants and virulence factors from several bacterial strains was monitored by multiplex PCR followed by hybridization of the denatured PCR product to gene-specific oligonucleotide probes on a microarray.[108] The results from this work suggest that microarray analysis of microbial virulence factors might be very useful for automated identification and characterization of bacterial pathogens.

A similar approach was used for other food pathogens including the analysis of the genes for the heat-stable enterotoxins (SEs), a family of 18 major serological types of toxins (SEA through SEV) representing one of the leading causes of gastroenteritis (vomiting and diarrhea) resulting from consumption of contaminated food. In this work, the microarray analysis demonstrated that many *S. aureus* strains contain multiple toxin genes[94] and that some of these isolates contain previously undetected enterotoxin genes. Other microarrays developed for the analysis of *B. anthracis* virulence factors include *pagA, lef* and *cya*,[74,81] *Listeria*,[83] *Campylobacter* species,[77] *Clostridium perfringens*[80,109] and discrimination between pathogenic 0157:H7 and non-pathogenic *Escherichia coli* strains.[110]

In addition to food analysis, microarrays were used for the detection of bacterial pathogens in municipal wastewater.[108,111,112] Beyond microbial diagnostics, microarrays were used for investigation of *Campylobacter* diversity and pathogenesis specific genes by whole microbial genome comparisons demonstrating the broad utility of the technology,[113] and for fingerprinting of *Bacillus* isolates.[114]

In terms of sensitivity of the method, it is useful to note that conventional methods have a detection limit of one cell per 25-g sample.[115] DNA microarrays with oligoprobes complementary to four *E. coli* O157:H7 virulence loci (intimin, Shiga-like toxins I and II, and hemolyxin A) were used to detect less than one cell equivalent of genomic DNA (1 fg)[116] using a genomic DNA microarray employing 10 functional genes as detection targets. Sensitivity of the microarray was determined to be approximately 1.0 µg of *Escherichia coli* genomic DNA, or 2×10^8 copies of the target gene. The sensitivity of the microarray was enhanced by approximately six orders of magnitude when the target 23S rRNA gene sequences were PCR amplified with a novel universal primer set and hybridized to 24 species-specific oligonucleotide probes. The minimum detection limit was estimated to be about 100 fg of *E. coli* genomic DNA or 1.4×10^2 copies of the 23S rRNA gene. The PCR amplified DNA microarray successfully detected multiple bacterial pathogens in wastewater.

6.7.1.2 Multi-pathogen Microarrays

In addition to microarray analysis of individual pathogens, several pathogen-specific microarrays were integrated into a single DNA chip[75] for simultaneous analysis of multiple virulence factors including *S. aureus* enterotoxin genes, *Listeria* spp., *Campylobacter* spp. and *Clostridium perfringens*, which represent the majority of food microbial pathogens. Similarly, genomic markers were used for a high-sensitivity pathogen detection microarray (10 fg of *B. anthracis*). Target sequences were PCR-amplified, targeting 18 potential biowarfare agents.[104] A larger microarray with over 53 000 oligoprobes was used for multi-pathogen (142 unique diagnostic regions of 11 bacteria, 5 RNA viruses and 2 eukaryotes) analysis.[103]

When analyzing multiple pathogens, the identification of pathogenic bacteria in a background of non-pathogens is a challenge, and especially the detection and identification of low-abundance pathogens within a complex microbial community. A microbial diagnostic microarray using single-nucleotide extension labeling with gyrB as the marker gene was used[102,117–119] for specific detection of a broad range of pathogenic bacteria. A microarray with 35 oligonucleotide probes[119] targeting *Escherichia coli*, *Shigella* spp., *Salmonella* spp., *Aeromonas hydrophila*, *Vibrio cholerae*, *Mycobacterium avium*, *Mycobacterium tuberculosis*, *Helicobacter pylori*, *Proteus mirabilis*, *Yersinia enterocolitica* and *Campylobacter jejuni* was developed. The introduction of competitive oligonucleotides in the labeling reaction successfully suppressed cross-reaction by closely related sequences, significantly improving the performance of the assay. Environmental performance was tested with environmental and veterinary samples harboring complex microbial communities. Detection sensitivity in the range of 0.1% has been demonstrated, far below the 5% detection limit of traditional microbial diagnostic microarrays. This trend of developing microarrays for multi-organism analysis opens new applications in the analysis of microbial communities.

6.7.1.3   Microarrays for Resequencing

Microarray technology was used for resequencing of multiple *Bacillus anthracis* isolates.[120] The array covered 3.1 Mb of genomic sequence from a panel of 56 *Bacillus anthracis* strains. Sequence quality was shown to be very high (discrepancy rate of 7.4×10^{-7}). Such microarray-based rapid resequencing technologies (resequencing arrays) may be critical for recognizing newly emerging or genetically engineered strains.

6.7.2   Microarray Analysis of Microbial Communities

The ability of a microarray to analyze many DNA sequences simultaneously makes the technology ideal for analysis of diverse microbial communities, including tracking of environmental isolates, measuring the dynamics of populations and environmental impacts on microbial communities. A program was developed (ProDesign) for design of microarrays to track gene families present in environmental samples.[42] Several microarray approaches were developed and used for microbial communities analysis including the following.

6.7.2.1   Functional Gene Arrays (FGA)

Functional Gene Arrays are used for detection of genes involved in specific functions such as biodegradation and biotransformation in microbial communities[121] and for analysis of genes involved in nitrogen cycling.[122]

6.7.2.2   Community Genome Arrays (CGA)

Community Gene Arrays are microarrays which use probes derived from whole genomic DNA isolated from a bacterial community. A CGA with DNA from 67 closely or distantly related representative bacterial strains revealed differences in microbial community composition in soil, river and marine sediments.[123]

6.7.2.3   Repetitive Sequences Microarray

Repetitive sequences that are interspersed throughout the genome of diverse bacterial species include the highly conserved 154-bp BOX palindromic sequence. A subset of the BOX sequence was used as a PCR primer[124] creating a "fingerprint" for each species. A random set of 198 9nt probes were used on a microarray for the purpose of assessing and managing the risk posed by microbial pollution. The microarray results enabled cluster hybridization profiles to be generated, which correlated with the environmental source from which the *Enterococcus* sp. isolates originated.

6.7.2.4   Random DNAs Representation Array

Another approach for screening DNA of unknown microbial communities is the "FloraArray".[125] This array was developed to represent the characteristics of a microbial community. Genomic DNA from a bacterial sludge sample was fragmented and the fragments were inserted into a vector to construct a shotgun library. The array was fabricated with 2000 random clones, from the library, as probes. DNA samples from the environment were analyzed by comparative hybridization on the array with a sample from a well-characterized anaerobic ammonium oxidation (anammox) bacterial community. The results demonstrated that the array provided ∼300 spots characteristic of the anammox community. It was concluded that this microarray-based approach has potential to be useful for analysis of unknown microbial communities.

6.7.2.5   Phylogenetic Oligonucleotide Arrays

As discussed above ribosomal RNA and corresponding genes (*rrn*) are used for bacterial identification and evolutionary classification determination leveraging genetic properties of this locus.[99] Several microarrays are based on ribosomal sequence used for microbial community analysis including rapid quantitative profiling of complex microbial populations analysis.[126] A DNA oligonucleotide microarray was presented composed of 10 462 small subunit (SSU) ribosomal DNA (rDNA) probes selected to provide quantitative information on the taxonomic composition of diverse microbial populations. The microarray enabled detection and quantification of individual bacterial species present at fractional abundances of <0.1% in complex synthetic mixtures.

Several examples employing ribosomal sequence-based microarrays for microbial community analysis include:

- Sulfate-reducing prokaryote[127] compost microbial communities[128] with detection limit of 10^5 cells, in compost spiking experiments

- Chemical-contaminated soils; the composition of microbial communities in hexachlorocyclohexane (HCH) contaminated soils from Spain revealed with a habitat-specific microarray[129]
- Freshwater sediments; a DNA microarray platform based on direct detection of rRNA for characterization of freshwater sediment-related prokaryotic communities[130]
- Uranium migration; application of a high-density oligonucleotide microarray approach to study bacterial population dynamics during uranium reduction and reoxidation[131]
- Microbial populations in acid mine drainage and bioleaching systems[132]
- Nickel-tolerant microorganisms from contaminated sediments[133]
- Air monitoring of urban aerosols[134]

6.7.2.6 Sensitivity of Microarray Analysis of Microbial Communities

Current detection sensitivities are often not sufficient for detecting the less dominant microbial populations in an environmental sample. For single-copy genes, genomic DNA from approximately 10^7 cells is required to obtain a reasonably strong microarray signal (using 50mer-based oligonucleotide microarrays).[121] Directly extracted rRNA from environmental microbial populations (from sediment cores), without PCR amplification, were analyzed with a microarray consisting of 21 oligonucleotide rRNA probes[135] yielding a detection limit of 0.5 µg of total rRNA (similar to the level of PCR detection limits applied to environmental systems (10^9 to 10^{10} copies of 16S rRNA)).

Multiple displacement amplification (MDA) was developed for a whole-community genome amplification (WCGA) microarray detection approach to analyze microbial community structure.[136] Very low concentrations of DNA (as low as 10 fg) could be detected, but the lower template concentrations affected the representativeness of the WCGA amplification. WCGA was used to investigate the microbial communities in groundwater contaminated with uranium and other metals.

6.7.3 Analysis of Antibiotic Resistance

Clinically it is very important to determine the antibiotic resistance profile of a contaminating organism to enable provision of the appropriate treatment to infected people and to enable monitoring of the spread of antibiotic-resistant microorganisms. Traditional antibiotic resistance determination is based on culturing, which may be slow for bacteria such as *Mycobacterium tuberculosis* and may not provide information about the mechanism of the resistance. Microarrays can provide molecular level information on antibiotic resistance and several microarrays were developed to provide such an analysis for bacteria, yeasts and other pathogenic microorganisms. In general the application of microarrays for analysis of antibiotics resistance can be divided into two main groups: 1) microarrays for diagnostic and genotyping of antibiotic resistance determinants and 2) microarrays for studying the mechanism of antibiotic resistance.

6.7.3.1   Microarrays for Diagnostic and Genotyping of Antibiotic Resistance Determinants

Several microarrays for analysis of *Staphylococcus aureus* antibiotic resistance were developed. To demonstrate the utility of the technology, a detailed example of a microarray for antibiotic resistance analysis is described.

A microarray for analysis of *Staphylococcus aureus* Erythromycin resistance determinants was developed and tested.[76] The microarray contains six 20–30 nt oligoprobes representing each of the six genes (*ermA, ermB, ermC, ereA, ereB, msrA/B*) that account for more than 98% of Erythromycin (and the related macrolide-lincosamide-streptogramin (MLS)) resistance. These are important antibiotics used for Gram-positive and some Gram-negative bacteria. The microarray was used to test reference and clinical *S. aureus* and *Streptococcus pyogenes* strains. Target genes from the samples were amplified and fluorescently labeled using multiplex PCR target amplification.

To simplify the initial step of preparing dye-labeled amplicons from each gene and to reduce the number of PCR reactions needed for MLS resistance analysis, two multiplex PCR reaction primer sets were used to amplify all six genes and the MLS microarray was used to determine which MLS resistance genes were present in the samples. An example microarray image from this analysis is shown in Figure 6.5, where each of the six genes analyzed is represented by seven different probes from different parts of the genes enabling the detection of various alleles of the genes. As shown in Figure 6.5, three of the strains contain multiple antibiotic resistance determinants. Out of 18 *S. aureus* clinical strains tested, 11 isolates carry MLS determinants. One gene (*ermC*) was found in all 11 clinical isolates tested, and 2 others, *ermA* and *msrA/B*, were found in 5 or more isolates. Indeed, 8 (72%) of the 11 clinical isolate strains contained 2 or 3 MLS resistance genes, in 1 of 3 combinations (*ermA* with *ermC, ermC* with *msrA/B, ermA* with *ermC* and *msrA/B*). These results demonstrate that microarray-based detection of microbial antibiotic resistance determinants provides considerable insight into the genetic profile of these aggressive pathogens which may be useful in designing therapy in the clinic and in designing strategies for public health.

For a similar application, a microarray was developed for the detection of 10 clinically and therapeutically relevant antibiotic resistance genes and mutations in *S. aureus* (mecA, aacA-aphD, tetK, tetM, vat(A), vat(B), vat(C), erm(A), erm(C), grlA-mutation).[137]

In other studies, multiple tetracycline (*tet*) resistance genes and ß-lactamase *bla*TEM-1 genes in *Escherichia coli* were assayed.[138] Another analysis[139] included multiple antimicrobial resistance genes, including *aadA, tetA* and *sulI*, which were most commonly detected in bacteria resistant to streptomycin, tetracycline

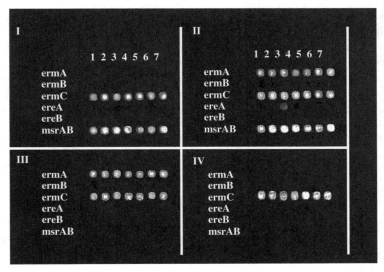

Figure 6.5 DNA microarray hybridization patterns of *S. aureus* erythromycin resistant clinical strains. Multiplex PCR amplification products of samples from four erythromycin-resistant clinical strains (I-IV) were hybridized to the microarray with six erythromycin resistance genes, each of the six genes analyzed is represented by seven different probes (numbered 1–7) from different parts of each gene. Panels: I – isolate M857; II – isolate M802; III – isolate M654; IV – isolate M655.

and sulfonamide. It also included the bla(CMY-2) and bla(TEM-1) genes, conferring resistance to third-generation cephalosporins in *Salmonella* and *E. coli*. In addition to genomic analysis, mutations which can cause the severe extended-spectrum beta-lactamase (ESBL) or inhibitor-resistant TEM (IRT) phenotype (causing resistance to extended-spectrum cephalosporins, mono-bactams and beta-lactamase inhibitors) were analyzed by SNP microarray for *Escherichia coli*, *Enterobacter cloacae* and *Klebsiella pneumoniae*.[140] Screening for resistance genes was performed by PCR using specific primers, or using a DNA microarray with around 300 nucleotide probes representing 7 classes of antibiotic resistance genes used for molecular characterization of intrinsic and acquired antibiotic resistance in lactic acid bacteria and bifidobacteria.[141] The genes identified encoded resistance to tetracycline (tet(M), tet(W), tet(O) and tet(O/W)), erythromycin and clindamycin [erm(B)] and streptomycin (aph(E) and sat(3)). Internal portions of some of these determinants were sequenced and found to be identical to genes described in other bacteria.

A disposable microarray was developed for detection of up to 90 antibiotic resistance genes in Gram-positive bacteria by hybridization[142] enabling the detection of multidrug-resistant strains of *Enterococcus faecalis*, *Enterococcus faecium*, *Lactococcus lactis*, an avirulent strain of *Bacillus anthracis* harboring the broad-host-range resistance plasmid, *Staphylococcus haemolyticus* and *Clostridium perfringens*. This technology has a large potential for applications in basic research, food safety and surveillance programs for antimicrobial resistance.

Determination of antibiotic resistance in *Mycobacterium tuberculosis* is challenging because of slow growth characteristic of this pathogen. A TB-Biochip oligonucleotide microarray was developed[143] as a rapid system to detect mutations associated with rifampin (RIF) resistance in mycobacteria with a sensitivity of 80% and a specificity of 100% relative to conventional drug susceptibility testing results for RIF resistance. Other microarrays were used for mapping of mutations of pyrazinamide-resistant *Mycobacterium tuberculosis* strains.[144]

6.7.3.2 Microarrays for Studying the Mechanism of Antibiotic Resistance

In several studies, microarrays were used to determine antibiotic resistance mechanisms and pathways, through changes in gene expression in response to environmental changes. These results all contribute to better understanding of drug resistance which may enable a solution to antibiotic resistance through development of predictive models in the area of antibiotic toxicogenomics and through the development of new drugs. Unlike diagnostic studies through genotyping of a few antibiotic resistance determinants, studying pathways often relies on genome-wide expression profiling with several mechanisms including:

- A microarray-based antibiotic screen used to study the regulatory role of supercoiling in the osmotic stress response of *Escherichia coli* exposed to novobiocin, pefloxacin and chloramphenicol.[145]
- Microarray transcription analysis of clinical *Staphylococcus aureus* isolates resistant to vancomycin identified 35 genes with increased transcription and 16 genes with decreased transcription,[146] many involved in purine biosynthesis or transport suggesting that increased energy (ATP) is required to generate the thicker cell walls that characterize resistant mutants. Microarray expression profiling of *Yersinia pestis* in response to chloramphenicol[147] identified 755 genes which were differentially expressed on chloramphenicol treatment: 364 genes were up-regulated and 391 were down-regulated. Genes encoding the components of the translation apparatus, cell envelope and transport/binding functions were strongly represented amongst the induced genes. Genes encoding proteins involved in energy metabolism and synthesis and modification of macromolecules were strongly represented amongst the down-regulated genes. Similarly, the global gene expression profile of *Yersinia pestis* induced by strepto-mycin identified 345 genes that were differentially regulated, 144 of which were up-regulated and 201 down-regulated. Streptomycin-induced tran-scriptional changes occurred in genes responsible for heat shock response, drug/analogue sensitivity, biosynthesis of the branched-chain amino acids, chemotaxis, mobility and broad regulatory functions.[148]

- Evaluation of differential gene expression in fluconazole susceptible and resistant isolates of *Candida albicans* (an opportunistic fungal pathogen causing oropharyngeal candidiasis (OPC) in AIDS) by microarray analysis identified genes which are differentially expressed in association with azole resistance.[149] These included genes involved in amino acid and carbohydrate metabolism; cell stress; cell-wall maintenance; lipid, fatty acid and sterol metabolism; and small molecule transport.

- Genome-wide expression profiling revealed genes associated with amphotericin B and fluconazole resistance.[72] This study identified 134 genes which were found to be differentially expressed. In addition to the cell stress genes, the ergosterol biosynthesis genes and several histone genes, protein synthesis genes and energy generation genes were down-regulated. The response of *Mycobacterium tuberculosis* to six antimicrobial agents was determined by microarray analysis in an attempt to define mechanisms of innate resistance in *M. tuberculosis*. The gene expression profiles of *M. tuberculosis* after treatment with several antibiotics established an expression profile which overlapped with a number of other mycobacterial stress responses and elucidated a novel pathway contributing to mycobacterial drug resistance.

6.7.4 Microarrays for Applications in Agricultural Settings

One example is the use of a microarray to study antibiotic susceptibility patterns and resistance genes of starter cultures and probiotic bacteria used in food. Isolates exhibiting resistance that is not an intrinsic feature of the respective genera were analyzed by microarray hybridization as a tool to trace phenotypic resistance to specific genetic determinants. This study resulted in the detection of several antibiotics resistance determinants in these cultures.[150]

A second example is a microarray for detection of antibiotic resistance genes of pathogenic *Salmonella* from swine. A microarray was developed to detect 11 antibiotic resistance genes that confer resistance to aminoglycosides, tetracyclines, sulfonamides and chloramphenicols.[151]

6.8 Summary

DNA microarrays, which were originally developed for gene expression analysis, show excellent potential as a tool for microbial genetic analysis in research, clinical, agricultural, regulatory, public health, industrial and ecological settings. Unlike microarray technology used for gene expression, which provides a platform for analysis of tens of thousands of parallel genetic determinants, the microarray genotyping analysis described here involves far smaller arrays which analyze only a subset of genes (*e.g.* virulence factors, mutations which confer antibiotics resistance, pathogencity islands or ribosomal genes). Another difference is that most of the microarrays described here are "homemade" in contrast to the large commercial microarrays developed for gene expression.

The technology available today, which includes relatively simple contact printers and inexpensive scanners, opens new opportunities, even to small laboratories, to develop new clinical and environmental applications for microarray technology. Such applications may move microbial identification and characterization forward from the traditional culture and immunological methodologies to a new era of genomics based methods.

Acknowledgements

This work was supported in part by USDA grant 200013000 to A.R. and USDA grant 20033520113784 to K.H., additional funding provided by the FDA Office of Science.

Acronyms

AIDS	Acquired Immune Deficiency Syndrome
aRNA	Antisense RNA
ATP	Adenosine Triphosphate
BLAST	Basic Local Alignment Search Tool
bp	Base pare
CCD	Charge-Coupled Device
cDNA	Complementary DNA
CGA	Community Genome Array
CGH	Comparative Genomic Hybridization
DNA	Deoxyribonucleic acid
DOP-PCR	Degenerated Oligonucleotide Primed PCR
ELISA	Enzyme-Linked Immunosorbent Assays
FDA	Food and Drug Administration
FGA	Functional Gene Array
GC	content Guanine-Cytosine content
Indels	Insertions/Deletions
LAPT	Linear Amplification of Prokaryotic Transcripts
LSU	Large-Subunit (rRNA)
MDA	Multiple Displacement Amplification
mer	(number of) nucleotides
MLS	Macrolide-Lincosamide-Streptogramin
mRNA	Messenger Ribonucleic Acid
NCBI	National Center for Biotechnology Information
nt	Nucleotides
oligo(dT)	Oligodeoxythymidylic acid
OPC	Oropharyngeal Candidiasis
PCR	Polymerase Chain Reaction

PMMA	Poly(methyl methacrylate (Acrylic)
PMT	Photomultiplier Tube
Poly(A)	Polyadenosine
Poly(T)	Polythymidine
RIF	Rifampin
RNA	Ribonucleic acid
rrn	Ribosomal Ribonucleic Acid
rRNA	Ribosomal DNA
RT-PCR	ReverseTranscription Polymerase Chain Reaction
SDS	Sodium Dodecyl Sulfate
SNP	Single Nucleotide Polymorphism
ssDNA	Single-Stranded DNA
ssRNA	Single-Stranded RNA
SSU	Small Subunit (RNA)
Tm	Melting Temperature
UV	Ultra Violate
WCGA	Whole-Community Genome Amplification
WGA	Whole Genome Amplification

References

1. E. Engvall and P. Perlman, Enzyme-linked immunosorbent assay (ELISA). Quantitative assay of immunoglobulin G, *Immunochemistry*, 1971, **8**(9), 871–874.

2. D. Gillespie and S. Spiegelman, A quantitative assay for DNA-RNA hybrids with DNA immobilized on a membrane, *J. Mol. Biol.*, 1965, **12**(3), 829–842.

3. J. G. Gall and M. L. Pardue, Formation and detection of RNA-DNA hybrid molecules in cytological preparations, *Proc. Natl. Acad. Sci. USA*, 1969, **63**(2), 378–383.

4. A. M. Ayulo, R. A. Machado and V. M. Scussel, Enterotoxigenic Escherichia coli and Staphylococcus aureus in fish and seafood from the southern region of Brazil, *Int. J. Food Microbiol.*, 1994, **24**(1–2), 171–178.

5. J. C. Alwine, D. J. Kemp and G. R. Stark, Method for detection of specific RNAs in agarose gels by transfer to diazobenzyloxymethyl-paper and hybridization with DNA probes, *Proc. Natl. Acad. Sci. USA*, 1977, **74**(12), 5350–5354.

6. F. C. Kafatos, C. W. Jones and A. Efstratiadis, Determination of nucleic acid sequence homologies and relative concentrations by a dot hybridization procedure, *Nucleic Acids Res.*, 1979, **7**(6), 1541–1552.

7. R. Ekins, F. Chu and J. Micallef, High specific activity chemiluminescent and fluorescent markers: their potential application to high sensitivity and 'multi-analyte' immunoassays, *J. Biolumin. Chemilumin.*, 1989, **4**(1), 59–78.

8. R. Ekins, F. Chu and E. Biggart, Multispot, multianalyte, immunoassay, *Ann. Biol. Clin. (Paris)*, 1990, **48**(9), 655–666.

9. R. Ekins and F. W. Chu, Microarrays: their origins and applications, *Trends Biotechnol.*, 1999, **17**(6), 217–218.
10. T. Degenkolbe, M. A. Hannah, S. Freund, D. K. Hincha, A. G. Heyer and K. I. Kohl, A quality-controlled microarray method for gene expression profiling, *Anal. Biochem.*, 2005, **346**(2), 217–224.
11. K. L. Gunderson, S. Kruglyak, M. S. Graige, F. Garcia, B. G. Kermani, C. Zhao, D. Che, T. Dickinson, E. Wickham, J. Bierle, D. Doucet, M. Milewski, R. Yang, C. Siegmund, J. Haas, L. Zhou, A. Oliphant, J. B. Fan, S. Barnard and M. S. Chee, Decoding randomly ordered DNA arrays, *Genome Res.*, 2004, **14**(5), 870–877.
12. Y. Kohara, H. Noda, K. Okano and H. Kambara, DNA hybridization using "bead-array": probe-attached beads arrayed in a capillary in a predetermined order, *Nucleic Acids Res. Suppl.*, **2001**(1), 83–84.
13. K. Kuhn, S. C. Baker, E. Chudin, M. H. Lieu, S. Oeser, H. Bennett, P. Rigault, D. Barker, T. K. McDaniel and M. S. Chee, A novel, high-performance random array platform for quantitative gene expression profiling, *Genome Res.*, 2004, **14**(11), 2347–2356.
14. A. Spiro, M. Lowe and D. Brown, A bead-based method for multiplexed identification and quantitation of DNA sequences using flow cytometry, *Appl. Environ. Microbiol.*, 2000, **66**(10), 4258–4265.
15. T. M. Straub, B. P. Dockendorff, M. D. Quinonez-Diaz, C. O. Valdez, J. I. Shutthanandan, B. J. Tarasevich, J. W. Grate and C. J. Bruckner-Lea, Automated methods for multiplexed pathogen detection, *J. Microbiol. Methods*, 2005, **62**(3), 303–316.
16. R. C. Summerbell, C. A. Levesque, K. A. Seifert, M. Bovers, J. W. Fell, M. R. Diaz, T. Boekhout, G. S. de Hoog, J. Stalpers and P. W. Crous, Microcoding: the second step in DNA barcoding, *Philos. Trans. R. Soc. Lond. B Biol. Sci.*, 2005, **360**(1462), 1897–1903.
17. D. Deregt, S. A. Gilbert, S. Dudas, J. Pasick, S. Baxi, K. M. Burton and M. K. Baxi, A multiplex DNA suspension microarray for simultaneous detection and differentiation of classical swine fever virus and other pestiviruses, *J. Virol. Methods*, 2006, **136**(1–2), 17–23.
18. M. R. Diaz, T. Boekhout, B. Theelen, M. Bovers, F. J. Cabanes and J. W. Fell, Microcoding and flow cytometry as a high-throughput fungal identification system for Malassezia species, *J. Med. Microbiol.*, 2006, **55**(Pt 9), 1197–1209.
19. P. Porschewski, M. A. Grattinger, K. Klenzke, A. Erpenbach, M. R. Blind and F. Schafer, Using aptamers as capture reagents in bead-based assay systems for diagnostics and hit identification, *J. Biomol. Screen.*, 2006, **11**(7), 773–781.
20. M. Schmitt, I. G. Bravo, P. J. Snijders, L. Gissmann, M. Pawlita and T. Waterboer, Bead-based multiplex genotyping of human papillomaviruses, *J. Clin. Microbiol.*, 2006, **44**(2), 504–512.
21. L. Yang, D. K. Tran and X. Wang, BADGE, Beads Array for the Detection of Gene Expression, a high-throughput diagnostic bioassay, *Genome Res.*, 2001, **11**(11), 1888–1898.

22. I. B. Baums, K. D. Goodwin, T. Kiesling, D. Wanless, M. R. Diaz and J. W. Fell, Luminex detection of fecal indicators in river samples, marine recreational water, and beach sand, *Mar. Pollut. Bull.*, 2007, **54**(5), 521–536.

23. B. T. Page, C. E. Shields, W. G. Merz and C. P. Kurtzman, Rapid identification of ascomycetous yeasts from clinical specimens by a molecular method based on flow cytometry and comparison with identifications from phenotypic assays, *J. Clin. Microbiol.*, 2006, **44**(9), 3167–3171.

24. B. T. Page and C. P. Kurtzman, Rapid identification of Candida species and other clinically important yeast species by flow cytometry, *J. Clin. Microbiol.*, 2005, **43**(9), 4507–4514.

25. J. SantaLucia Jr., A unified view of polymer, dumbbell, and oligonucleotide DNA nearest-neighbor thermodynamics, *Proc. Natl. Acad. Sci. USA*, 1998, **95**(4), 1460–1465.

26. N. Sugimoto, S. Nakano, M. Yoneyama and K. Honda, Improved thermodynamic parameters and helix initiation factor to predict stability of DNA duplexes, *Nucleic Acids Res.*, 1996, **24**(22), 4501–4505.

27. K. J. Breslauer, R. Frank, H. Blocker and L. A. Marky, Predicting DNA duplex stability from the base sequence, *Proc. Natl. Acad. Sci. USA*, 1986, **83**(11), 3746–3750.

28. K. E. Herold and A. Rasooly, Oligo Design: a computer program for development of probes for oligonucleotide microarrays, *Biotechniques*, 2003, **35**(6), 1216–1221.

29. L. Bodrossy and A. Sessitsch, Oligonucleotide microarrays in microbial diagnostics, *Curr. Opin. Microbiol.*, 2004, **7**(3), 245–254.

30. Z. He, L. Wu, M. W. Fields and J. Zhou, Use of microarrays with different probe sizes for monitoring gene expression, *Appl. Environ. Microbiol.*, 2005, **71**(9), 5154–5162.

31. J. G. Hacia, Resequencing and mutational analysis using oligonucleotide microarrays, *Nat. Genet.*, 1999, **21**(1 Suppl), 42–47.

32. J. G. Hacia and F. S. Collins, Mutational analysis using oligonucleotide microarrays, *J. Med. Genet.*, 1999, **36**(10), 730–736.

33. J. G. Hacia, J. B. Fan, O. Ryder, L. Jin, K. Edgemon, G. Ghandour, R. A. Mayer, B. Sun, L. Hsie, C. M. Robbins, L. C. Brody, D. Wang, E. S. Lander, R. Lipshutz, S. P. Fodor and F. S. Collins, Determination of ancestral alleles for human single-nucleotide polymorphisms using high-density oligonucleotide arrays, *Nat. Genet.*, 1999, **22**(2), 164–167.

34. O. P. Kallioniemi, Biochip technologies in cancer research, *Ann. Med.*, 2001, **33**(2), 142–147.

35. D. J. Cutler, M. E. Zwick, M. M. Carrasquillo, C. T. Yohn, K. P. Tobin, C. Kashuk, D. J. Mathews, N. A. Shah, E. E. Eichler, J. A. Warrington and A. Chakravarti, High-throughput variation detection and genotyping using microarrays, *Genome Res.*, 2001, **11**(11), 1913–1925.

36. S. F. Altschul, W. Gish, W. Miller, E. W. Myers and D. J. Lipman, Basic local alignment search tool, *J. Mol. Biol.*, 1990, **215**(3), 403–410.

37. D. L. Leiske, A. Karimpour-Fard, P. S. Hume, B. D. Fairbanks and R. T. Gill, A comparison of alternative 60-mer probe designs in an in-situ synthesized oligonucleotide microarray, *BMC Genomics, Paper 72*, 2006, **7**, 1–9.

38. Z. Bozdech, J. Zhu, M. P. Joachimiak, F. E. Cohen, B. Pulliam and J. L. DeRisi, Expression profiling of the schizont and trophozoite stages of Plasmodium falciparum with a long-oligonucleotide microarray, *Genome Biol. Paper R9*, 2003, **4**(2), 1–15.

39. M. Kuroda, T. Ohta, I. Uchiyama, T. Baba, H. Yuzawa, I. Kobayashi, L. Cui, A. Oguchi, K. Aoki, Y. Nagai, J. Lian, T. Ito, M. Kanamori, H. Matsumaru, A. Maruyama, H. Murakami, A. Hosoyama, Y. Mizutani-Ui, N. K. Takahashi, T. Sawano, R. Inoue, C. Kaito, K. Sekimizu, H. Hirakawa, S. Kuhara, S. Goto, J. Yabuzaki, M. Kanehisa, A. Yamashita, K. Oshima, K. Furuya, C. Yoshino, T. Shiba, M. Hattori, N. Ogasawara, H. Hayashi and K. Hiramatsu, Whole genome sequencing of meticillin-resistant Staphylococcus aureus, *Lancet*, 2001, **357**(9264), 1225–1240.

40. E. Talla, F. Tekaia, L. Brino and B. Dujon, A novel design of whole-genome microarray probes for Saccharomyces cerevisiae which minimizes cross-hybridization, *BMC Genomics*, 2003, **4**(1), 38.

41. Y. Charbonnier, B. Gettler, P. Francois, M. Bento, A. Renzoni, P. Vaudaux, W. Schlegel and J. Schrenzel, A generic approach for the design of whole-genome oligoarrays, validated for genomotyping, deletion mapping and gene expression analysis on Staphylococcus aureus, *BMC Genomics*, 2005, **6**, 95.

42. S. Feng and E. R. Tillier, A fast and flexible approach to oligonucleotide probe design for genomes and gene families, *Bioinformatics*, 2007, **23**(10), 1195–1202.

43. Y. H. Rogers, P. Jiang-Baucom, Z. J. Huang, V. Bogdanov, S. Anderson and M. T. Boyce-Jacino, Immobilization of oligonucleotides onto a glass support via disulfide bonds: A method for preparation of DNA microarrays, *Anal. Biochem.*, 1999, **266**(1), 23–30.

44. M. Beier and J. D. Hoheisel, Versatile derivatisation of solid support media for covalent bonding on DNA-microchips, *Nucleic Acids Res.*, 1999, **27**(9), 1970–1977.

45. L. A. Chrisey, C. E. O'Ferrall, B. J. Spargo, C. S. Dulcey and J. M. Calvert, Fabrication of patterned DNA surfaces, *Nucleic Acids Res.*, 1996, **24**(15), 3040–3047.

46. A. Csaki, R. Moller, W. Straube, J. M. Kohler and W. Fritzsche, DNA monolayer on gold substrates characterized by nanoparticle labeling and scanning force microscopy, *Nucleic Acids Res.*, 2001, **29**(16), E81.

47. B. G. Healey, R. S. Matson and D. R. Walt, Fiberoptic DNA sensor array capable of detecting point mutations, *Anal. Biochem.*, 1997, **251**(2), 270–279.

48. F. Fixe, M. Dufva, P. Telleman and C. B. Christensen, One-step immobilization of aminated and thiolated DNA onto poly(methylmethacrylate) (PMMA) substrates, *Lab Chip*, 2004, **4**(3), 191–195.

49. S. A. Soper, M. Hashimoto, C. Situma, M. C. Murphy, R. L. McCarley, Y. W. Cheng and F. Barany, Fabrication of DNA microarrays onto polymer substrates using UV modification protocols with integration into microfluidic platforms for the sensing of low-abundant DNA point mutations, *Methods*, 2005, **37**(1), 103–113.

50. F. Xu, P. Datta, H. Wang, S. Gurung, M. Hashimoto, S. Wei, J. Goettert, R. L. McCarley and S. A. Soper, Polymer Microfluidic Chips with Integrated Waveguides for Reading Microarrays, *Anal. Chem.*, 2007, **79**(23), 9007–9013.

51. H. Mao, H. Wang, D. Zhang, H. Mao, J. Zhao, J. Shi and Z. Cui, Study of hepatitis B virus gene mutations with enzymatic colorimetry-based DNA microarray, *Clin. Biochem.*, 2006, **39**(1), 67–73.

52. B. A. Stillman and J. L. Tonkinson, FAST slides: a novel surface for microarrays, *Biotechniques*, 2000, **29**(3), 630–635.

53. R. F. Wang, S. J. Kim, L. H. Robertson and C. E. Cerniglia, Development of a membrane-array method for the detection of human intestinal bacteria in fecal samples, *Mol. Cell. Probes*, 2002, **16**(5), 341–350.

54. G. Wrobel, J. Schlingemann, L. Hummerich, H. Kramer, P. Lichter and M. Hahn, Optimization of high-density cDNA-microarray protocols by 'design of experiments', *Nucleic Acids Res.*, 2003, **31**(12), e67.

55. D. Guschin, G. Yershov, A. Zaslavsky, A. Gemmell, V. Shick, D. Proudnikov, P. Arenkov and A. Mirzabekov, Manual manufacturing of oligonucleotide, DNA, and protein microchips, *Anal. Biochem.*, 1997, **250**(2), 203–211.

56. M. Schena, D. Shalon, R. W. Davis and P. O. Brown, Quantitative monitoring of gene expression patterns with a complementary DNA microarray, *Science*, 1995, **270**(5235), 467–470.

57. P. Hegde, R. Qi, K. Abernathy, C. Gay, S. Dharap, R. Gaspard, J. E. Hughes, E. Snesrud, N. Lee, and J. Quackenbush, A concise guide to cDNA microarray analysis. *Biotechniques*, 2000, **29**(3): 548–50, 552–4, 556 passim.

58. A. L. Lloyd, B. J. Marshall and B. J. Mee, Identifying cloned Helicobacter pylori promoters by primer extension using a FAM-labelled primer and GeneScan analysis, *J. Microbiol. Methods.*, 2005, **60**(3), 291–298.

59. R. N. Van Gelder, M. E. von Zastrow, A. Yool, W. C. Dement, J. D. Barchas and J. H. Eberwine, Amplified RNA synthesized from limited quantities of heterogeneous cDNA, *Proc. Natl. Acad. Sci. USA*, 1990, **87**(5), 1663–1667.

60. N. N. Iscove, M. Barbara, M. Gu, M. Gibson, C. Modi and N. Winegarden, Representation is faithfully preserved in global cDNA amplified exponentially from sub-picogram quantities of mRNA, *Nat. Biotechnol.*, 2002, **20**(9), 940–943.

61. J. N. Lawson and S. A. Johnston, Amplification of sense-stranded prokaryotic RNA, *DNA Cell Biol.*, 2006, **25**(11), 627–634.

62. A. Badiee, H. G. Eiken, V. M. Steen and R. Lovlie, Evaluation of five different cDNA labeling methods for microarrays using spike controls, *BMC Biotechnol.*, 2003, **3**(1), 23.

63. S. Goswami, W. Wang, J. B. Wyckoff and J. S. Condeelis, Breast cancer cells isolated by chemotaxis from primary tumors show increased survival and resistance to chemotherapy, *Cancer Res.*, 2004, **64**(21), 7664–7667.

64. W. Lian, S. A. Litherland, H. Badrane, W. Tan, D. Wu, H. V. Baker, P. A. Gulig, D. V. Lim and S. Jin, Ultrasensitive detection of biomolecules with fluorescent dye-doped nanoparticles, *Anal. Biochem.*, 2004, **334**(1), 135–144.

65. L. A. Lyon, M. D. Musick and M. J. Natan, Colloidal Au-enhanced surface plasmon resonance immunosensing, *Anal. Chem.*, 1998, **70**(24), 5177–5183.

66. X. Pang, D. Zhou, Y. Song, D. Pei, J. Wang, Z. Guo and R. Yang, Bacterial mRNA purification by magnetic capture-hybridization method, *Microbiol. Immunol.*, 2004, **48**(2), 91–96.

67. F. Di Cello, Y. Xie, M. Paul-Satyaseela and K. S. Kim, Approaches to bacterial RNA isolation and purification for microarray analysis of Escherichia coli K1 interaction with human brain microvascular endo-thelial cells, *J. Clin. Microbiol.*, 2005, **43**(8), 4197–4199.

68. S. Barbaux, O. Poirier and F. Cambien, Use of degenerate oligonucleotide primed PCR (DOP-PCR) for the genotyping of low-concentration DNA samples, *J. Mol. Med.*, 2001, **79**(5–6), 329–332.

69. J. G. Paez, M. Lin, R. Beroukhim, J. C. Lee, X. Zhao, D. J. Richter, S. Gabriel, P. Herman, H. Sasaki, D. Altshuler, C. Li, M. Meyerson and W. R. Sellers, Genome coverage and sequence fidelity of phi29 poly-merase-based multiple strand displacement whole genome amplification, *Nucleic Acids Res.*, 2004, **32**(9), e71.

70. R. Luthra and L. J. Medeiros, Isothermal multiple displacement ampli-fication: a highly reliable approach for generating unlimited high mole-cular weight genomic DNA from clinical specimens, *J. Mol. Diagn.*, 2004, **6**(3), 236–242.

71. V. Gadkar and M. C. Rillig, Application of Phi29 DNA polymerase mediated whole genome amplification on single spores of arbuscular mycorrhizal (AM) fungi, *FEMS Microbiol Lett.*, 2005, **242**(1), 65–71.

72. K. S. Barker, S. Crisp, N. Wiederhold, R. E. Lewis, B. Bareither, J. Eckstein, R. Barbuch, M. Bard and P. D. Rogers, Genome-wide expression profiling reveals genes associated with amphotericin B and fluconazole resistance in experimentally induced antifungal resistant iso-lates of Candida albicans, *J. Antimicrob. Chemother.*, 2004, **54**(2), 376–385.

73. D. L. Barker, M. S. Hansen, A. F. Faruqi, D. Giannola, O. R. Irsula, R. S. Lasken, M. Latterich, V. Makarov, A. Oliphant, J. H. Pinter, R. Shen, I. Sleptsova, W. Ziehler and E. Lai, Two methods of whole-genome amplification enable accurate genotyping across a 2320-SNP linkage panel, *Genome Res.*, 2004, **14**(5), 901–907.

74. N. Sergeev, M. Distler, M. Vargas, V. Chizhikov, K. E. Herold and A. Rasooly, Microarray analysis of Bacillus cereus group virulence fac-tors, *J. Microbiol. Methods*, 2006a, **65**(3), 488–502.

75. N. Sergeev, M. Distler, S. Courtney, S. F. Al-Khaldi, D. Volokhov, V. Chizhikov and A. Rasooly, Multipathogen oligonucleotide microarray for environmental and biodefense applications, *Biosens. Bioelectron.*, 2004a, **20**(4), 684–698.
76. D. Volokhov, V. Chizhikov, K. Chumakov and A. Rasooly, Microarray analysis of erythromycin resistance determinants, *J. Appl. Microbiol.*, 2003, **95**(4), 787–798.
77. D. Volokhov, V. Chizhikov, K. Chumakov and A. Rasooly, Microarray-based identification of thermophilic Campylobacter jejuni, C. coli, C. lari, and C. upsaliensis, *J. Clin. Microbiol.*, 2003, **41**(9), 4071–4080.
78. S. F. Al-Khaldi, S. A. Martin, A. Rasooly and J. D. Evans, DNA microarray technology used for studying foodborne pathogens and microbial habitats: minireview, *J. AOAC Int.*, 2002, **85**(4), 906–910.
79. V. Chizhikov, A. Rasooly, K. Chumakov and D. D. Levy, Microarray analysis of microbial virulence factors, *Appl. Environ. Microbiol.*, 2001, **67**(7), 3258–3263.
80. S. F. Al-Khaldi, D. Villanueva and V. Chizhikov, Identification and characterization of Clostridium perfringens using single target DNA microarray chip, *Int. J. Food Microbiol.*, 2004, **91**(3), 289–296.
81. D. Volokhov, A. Pomerantsev, V. Kivovich, A. Rasooly and V. Chizhikov, Identification of Bacillus anthracis by multiprobe microarray hybridization, *Diagn. Microbiol. Infect. Dis.*, 2004, **49**(3), 163–171.
82. J. Johnson, K. Jinneman, G. Stelma, B. G. Smith, D. Lye, J. Messer, J. Ulaszek, L. Evsen, S. Gendel, R. W. Bennett, B. Swaminathan, J. Pruckler, A. Steigerwalt, S. Kathariou, S. Yildirim, D. Volokhov, A. Rasooly, V. Chizhikov, M. Wiedmann, E. Fortes, R. E. Duvall and A. D. Hitchins, Natural atypical Listeria innocua strains with Listeria monocytogenes pathogenicity island 1 genes, *Appl. Environ. Microbiol.*, 2004, **70**(7), 4256–4266.
83. D. Volokhov, A. Rasooly, K. Chumakov and V. Chizhikov, Identification of Listeria species by microarray-based assay, *J. Clin. Microbiol.*, 2002, **40**(12), 4720–4728.
84. X. Tang, S. L. Morris, J. J. Langone and L. E. Bockstahler, Microarray and allele specific PCR detection of point mutations in Mycobacterium tuberculosis genes associated with drug resistance, *J. Microbiol. Methods*, 2005, **63**(3), 318–330.
85. B. A. Stillman and J. L. Tonkinson, Expression microarray hybridization kinetics depend on length of the immobilized DNA but are independent of immobilization substrate, *Anal. Biochem.*, 2001, **295**(2), 149–157.
86. D. S. Dandy, P. Wu and D. W. Grainger, Array feature size influences nucleic acid surface capture in DNA microarrays, *Proc. Natl. Acad. Sci. USA*, 2007, **104**(20), 8223–8228.
87. Y. Sun, K. B. Jacobson and V. Golovlev, Label-free detection of biomolecules on microarrays using surface-colloid interaction, *Anal. Biochem.*, 2007, **361**(2), 244–252.

88. H. S. Lai, Y. Chen, W. H. Lin, C. N. Chen, H. C. Wu, C. J. Chang, P. H. Lee, K. J. Chang and W. J. Chen, Quantitative gene expression analysis by cDNA microarray during liver regeneration after partial hepatectomy in rats, *Surg. Today*, 2005, **35**(5), 396–403.
89. J. Petersen, M. Stangegaard, H. Birgens and M. Dufva, Detection of mutations in the beta-globin gene by colorimetric staining of DNA microarrays visualized by a flatbed scanner, *Anal. Biochem.*, 2007, **360**(1), 169–171.
90. S. Raychaudhuri, P. D. Sutphin, J. T. Chang and R. B. Altman, Basic microarray analysis: grouping and feature reduction, *Trends Biotechnol.*, 2001, **19**(5), 189–193.
91. A. K. Jarvinen, S. Hautaniemi, H. Edgren, P. Auvinen, J. Saarela, O. P. Kallioniemi and O. Monni, Are data from different gene expression microarray platforms comparable? *Genomics*, 2004, **83**(6), 1164–1168.
92. C. L. Yauk, M. L. Berndt, A. Williams and G. R. Douglas, Comprehensive comparison of six microarray technologies, *Nucleic Acids Res.*, 2004, **32**(15), e124.
93. Z. P. Aguilar, W. R. Vandaveer and I. Fritsch, Self-contained microelectrochemical immunoassay for small volumes using mouse IgG as a model system, *Analytical Chemistry*, 2002, **74**(14), 3321–3329.
94. N. Sergeev, D. Volokhov, V. Chizhikov and A. Rasooly, Simultaneous analysis of multiple staphylococcal enterotoxin genes by an oligonucleotide microarray assay, *J. Clin. Microbiol.*, 2004b, **42**(5), 2134–2143.
95. R. S. Gupta and E. Griffiths, Critical issues in bacterial phylogeny, *Theor. Popul. Biol.*, 2002, **61**(4), 423–434.
96. R. S. Gupta, The phylogeny and signature sequences characteristics of Fibrobacteres, Chlorobi, and Bacteroidetes, *Crit. Rev. Microbiol.*, 2004, **30**(2), 123–143.
97. S. Karlin and L. Brocchieri, Heat shock protein 70 family: multiple sequence comparisons, function, and evolution, *J. Mol. Evol.*, 1998, **47**(5), 565–577.
98. R. S. Gupta and V. Johari, Signature sequences in diverse proteins provide evidence of a close evolutionary relationship between the Deinococcus-thermus group and cyanobacteria, *J. Mol. Evol.*, 1998, **46**(6), 716–720.
99. G. J. Olsen, D. J. Lane, S. J. Giovannoni, N. R. Pace and D. A. Stahl, Microbial ecology and evolution: a ribosomal RNA approach, *Annu. Rev. Microbiol.*, 1986, **40**, 337–365.
100. W. Ludwig and K. H. Schleifer, Bacterial phylogeny based on 16S and 23S rRNA sequence analysis, *FEMS Microbiol. Rev.*, 1994, **15**(2–3), 155–173.
101. R. M. Anthony, T. J. Brown and G. L. French, Rapid diagnosis of bacteremia by universal amplification of 23S ribosomal DNA followed by hybridization to an oligonucleotide array, *J. Clin. Microbiol.*, 2000, **38**(2), 781–788.

102. K. Kakinuma, M. Fukushima and R. Kawaguchi, Detection and identification of Escherichia coli, Shigella, and Salmonella by microarrays using the gyrB gene, *Biotechnol. Bioeng.*, 2003, **83**(6), 721–728.

103. K. H. Wilson, W. J. Wilson, J. L. Radosevich, T. Z. DeSantis, V. S. Viswanathan, T. A. Kuczmarski and G. L. Andersen, High-density microarray of small-subunit ribosomal DNA probes, *Appl. Environ. Microbiol.*, 2002a, **68**(5), 2535–2541.

104. W. J. Wilson, C. L. Strout, T. Z. DeSantis, J. L. Stilwell, A. V. Carrano and G. L. Andersen, Sequence-specific identification of 18 pathogenic microorganisms using microarray technology, *Mol. Cell Probes*, 2002b, **16**(2), 119–127.

105. G. Mitterer, M. Huber, E. Leidinger, C. Kirisits, W. Lubitz, M. W. Mueller and W. M. Schmidt, Microarray-based identification of bacteria in clinical samples by solid-phase PCR amplification of 23S ribosomal DNA sequences, *J. Clin. Microbiol.*, 2004, **42**(3), 1048–1057.

106. U. Nubel, P. M. Schmidt, E. Reiss, F. Bier, W. Beyer and D. Naumann, Oligonucleotide microarray for identification of Bacillus anthracis based on intergenic transcribed spacers in ribosomal DNA, *FEMS Microbiol. Lett.*, 2004, **240**(2), 215–223.

107. G. Keramas, D. D. Bang, M. Lund, M. Madsen, S. E. Rasmussen, H. Bunkenborg, P. Telleman and C. B. Christensen, Development of a sensitive DNA microarray suitable for rapid detection of Campylobacter spp, *Mol. Cell Probes*, 2003, **17**(4), 187–196.

108. A. Loy, C. Schulz, S. Lucker, A. Schopfer-Wendels, K. Stoecker, C. Baranyi, A. Lehner and M. Wagner, 16S rRNA gene-based oligonucleotide microarray for environmental monitoring of the betaproteobacterial order "Rhodocyclales", *Appl. Environ. Microbiol.*, 2005, **71**(3), 1373–1386.

109. N. Sergeev, D. Volokhov, V. Chizhikov and A. Rasooly, Simultaneous analysis of multiple staphylococcal enterotoxin genes by an oligonucleotide microarray assay, *J. Clin. Microbiol.*, 2004, **42**(5), 2134–2143.

110. S. F. Al-Khaldi, K. M. Myers, A. Rasooly and V. Chizhikov, Genotyping of Clostridium perfringens toxins using multiple oligonucleotide microarray hybridization, *Mol. Cell Probes*, 2004, **18**(6), 359–367.

111. C. F. Wu, J. J. Valdes, W. E. Bentley and J. W. Sekowski, DNA microarray for discrimination between pathogenic 0157:H7 EDL933 and non-pathogenic Escherichia coli strains, *Biosens. Bioelectron.*, 2003, **19**(1), 1–8.

112. D. Y. Lee, K. Shannon and L. A. Beaudette, Detection of bacterial pathogens in municipal wastewater using an oligonucleotide microarray and real-time quantitative PCR, *J. Microbiol. Methods*, 2006, **65**(3), 453–467.

113. K. Lemarchand, F. Berthiaume, C. Maynard, J. Harel, P. Payment, P. Bayardelle, L. Masson and R. Brousseau, Optimization of microbial DNA extraction and purification from raw wastewater samples for downstream pathogen detection by microarrays, *J. Microbiol. Methods*, 2005, **63**(2), 115–126.

114. F. Poly, D. Threadgill and A. Stintzi, Identification of Campylobacter jejuni ATCC 43431-specific genes by whole microbial genome comparisons, *J. Bacteriol.*, 2004, **186**(14), 4781–4795.

115. D. P. Chandler, O. Alferov, B. Chernov, D. S. Daly, J. Golova, A. Perov, M. Protic, R. Robison, M. Schipma, A. White and A. Willse, Diagnostic oligonucleotide microarray fingerprinting of Bacillus isolates, *J. Clin. Microbiol.*, 2006, **44**(1), 244–250.

116. Y. Liu-Stratton, S. Roy and C. K. Sen, DNA microarray technology in nutraceutical and food safety, *Toxicol. Lett.*, 2004, **150**(1), 29–42.

117. D. R. Call, F. J. Brockman and D. P. Chandler, Detecting and genotyping Escherichia coli O157:H7 using multiplexed PCR and nucleic acid microarrays, *Int. J. Food Microbiol.*, 2001, **67**(1–2), 71–80.

118. M. Fukushima, K. Kakinuma, H. Hayashi, H. Nagai, K. Ito and R. Kawaguchi, Detection and identification of Mycobacterium species isolates by DNA microarray, *J. Clin. Microbiol.*, 2003, **41**(6), 2605–2615.

119. S. F. Gonzalez, M. J. Krug, M. E. Nielsen, Y. Santos and D. R. Call, Simultaneous detection of marine fish pathogens by using multiplex PCR and a DNA microarray, *J. Clin. Microbiol.*, 2004, **42**(4), 1414–1419.

120. T. Kostic, A. Weilharter, S. Rubino, G. Delogu, S. Uzzau, K. Rudi, A. Sessitsch and L. Bodrossy, A microbial diagnostic microarray technique for the sensitive detection and identification of pathogenic bacteria in a background of nonpathogens, *Anal. Biochem.*, 2007, **360**(2), 244–254.

121. M. E. Zwick, F. McAfee, D. J. Cutler, T. D. Read, J. Ravel, G. R. Bowman, D. R. Galloway and A. Mateczun, Microarray-based resequencing of multiple Bacillus anthracis isolates, *Genome Biol. Paper R10*, 2005, **6**(1), 1–13.

122. S. K. Rhee, X. Liu, L. Wu, S. C. Chong, X. Wan and J. Zhou, Detection of genes involved in biodegradation and biotransformation in microbial communities by using 50-mer oligonucleotide microarrays, *Appl. Environ. Microbiol.*, 2004, **70**(7), 4303–4317.

123. L. Wu, D. K. Thompson, G. Li, R. A. Hurt, J. M. Tiedje and J. Zhou, Development and evaluation of functional gene arrays for detection of selected genes in the environment, *Appl. Environ. Microbiol.*, 2001, **67**(12), 5780–5790.

124. L. Wu, D. K. Thompson, X. Liu, M. W. Fields, C. E. Bagwell, J. M. Tiedje and J. Zhou, Development and evaluation of microarray-based whole-genome hybridization for detection of microorganisms within the context of environmental applications, *Environ. Sci. Technol.*, 2004, **38**(24), 6775–6782.

125. K. J. Indest, K. Betts and J. S. Furey, Application of oligonucleotide microarrays for bacterial source tracking of environmental Enterococcus sp. isolates, *Int. J. Environ. Res. Public Health*, 2005, **2**(1), 175–185.

126. T. Yokoi, Y. Kaku, H. Suzuki, M. Ohta, H. Ikuta, K. Isaka, T. Sumino and M. Wagatsuma, 'FloraArray' for screening of specific DNA probes

representing the characteristics of a certain microbial community, *FEMS Microbiol. Lett.*, 2007, **273**(2), 166–171.

127. C. Palmer, E. M. Bik, M. B. Eisen, P. B. Eckburg, T. R. Sana, P. K. Wolber, D. A. Relman and P. O. Brown, Rapid quantitative profiling of complex microbial populations, *Nucleic Acids Res. Paper e5*, 2006, **34**(1), 1–10.

128. A. Loy, A. Lehner, N. Lee, J. Adamczyk, H. Meier, J. Ernst, K. H. Schleifer and M. Wagner, Oligonucleotide microarray for 16S rRNA gene-based detection of all recognized lineages of sulfate-reducing prokaryotes in the environment, *Appl. Environ. Microbiol.*, 2002, **68**(10), 5064–5081.

129. I. H. Franke-Whittle, S. H. Klammer and H. Insam, Design and application of an oligonucleotide microarray for the investigation of compost microbial communities, *J. Microbiol. Methods*, 2005, **62**(1), 37–56.

130. J. D. Neufeld, W. W. Mohn and V. de Lorenzo, Composition of microbial communities in hexachlorocyclohexane (HCH) contaminated soils from Spain revealed with a habitat-specific microarray, *Environ. Microbiol.*, 2006, **8**(1), 126–140.

131. J. Peplies, C. Lachmund, F. O. Glockner and W. Manz, A DNA microarray platform based on direct detection of rRNA for characterization of freshwater sediment-related prokaryotic communities, *Appl. Environ. Microbiol.*, 2006, **72**(7), 4829–4838.

132. E. L. Brodie, T. Z. Desantis, D. C. Joyner, S. M. Baek, J. T. Larsen, G. L. Andersen, T. C. Hazen, P. M. Richardson, D. J. Herman, T. K. Tokunaga, J. M. Wan and M. K. Firestone, Application of a high-density oligonucleotide microarray approach to study bacterial population dynamics during uranium reduction and reoxidation, *Appl. Environ. Microbiol.*, 2006, **72**(9), 6288–6298.

133. H. Yin, L. Cao, G. Qiu, D. Wang, L. Kellogg, J. Zhou, Z. Dai and X. Liu, Development and evaluation of 50-mer oligonucleotide arrays for detecting microbial populations in Acid Mine Drainages and bioleaching systems, *J. Microbiol. Methods*, 2007, **70**(1), 165–178.

134. J. D. Van Nostrand, T. V. Khijniak, T. J. Gentry, M. T. Novak, A. G. Sowder, J. Z. Zhou, P. M. Bertsch and P. J. Morris, Isolation and characterization of four gram-positive nickel-tolerant microorganisms from contaminated sediments, *Microbiol. Ecol.*, 2007, **53**(4), 670–682.

135. E. L. Brodie, T. Z. DeSantis, J. P. Parker, I. X. Zubietta, Y. M. Piceno and G. L. Andersen, Urban aerosols harbor diverse and dynamic bacterial populations, *Proc. Natl. Acad. Sci. USA*, 2007, **104**(1), 299–304.

136. S. El Fantroussi, H. Urakawa, A. E. Bernhard, J. J. Kelly, P. A. Noble, H. Smidt, G. M. Yershov and D. A. Stahl, Direct profiling of environmental microbial populations by thermal dissociation analysis of native rRNAs hybridized to oligonucleotide microarrays, *Appl. Environ. Microbiol.*, 2003, **69**(4), 2377–2382.

137. L. Wu, X. Liu, C. W. Schadt and J. Zhou, Microarray-based analysis of subnanogram quantities of microbial community DNAs by using

whole-community genome amplification, *Appl. Environ. Microbiol.*, 2006, **72**(7), 4931–4941.

138. B. Strommenger, C. Schmidt, G. Werner, B. Roessle-Lorch, T. T. Bachmann and W. Witte, DNA microarray for the detection of therapeutically relevant antibiotic resistance determinants in clinical isolates of Staphylococcus aureus, *Mol. Cell Probes*, 2007, **21**(3), 161–170.
139. D. R. Call, M. K. Bakko, M. J. Krug and M. C. Roberts, Identifying antimicrobial resistance genes with DNA microarrays, *Antimicrob. Agents Chemother.*, 2003, **47**(10), 3290–3295.
140. S. Chen, S. Zhao, P. F. McDermott, C. M. Schroeder, D. G. White and J. Meng, A DNA microarray for identification of virulence and antimicrobial resistance genes in Salmonella serovars and Escherichia coli, *Mol. Cell Probes*, 2005, **19**(3), 195–201.
141. V. Grimm, S. Ezaki, M. Susa, C. Knabbe, R. D. Schmid and T. T. Bachmann, Use of DNA microarrays for rapid genotyping of TEM beta-lactamases that confer resistance, *J. Clin. Microbiol.*, 2004, **42**(8), 3766–3774.
142. M. S. Ammor, A. B. Florez, A. H. van Hoek, C. G. de Los Reyes-Gavilan, H. J. Aarts, A. Margolles and B. Mayo, Molecular characterization of intrinsic and acquired antibiotic resistance in lactic Acid bacteria and bifidobacteria, *J. Mol. Microbiol. Biotechnol.*, 2008, **14**(1–3), 6–15.
143. V. Perreten, L. Vorlet-Fawer, P. Slickers, R. Ehricht, P. Kuhnert and J. Frey, Microarray-based detection of 90 antibiotic resistance genes of gram-positive bacteria, *J. Clin. Microbiol.*, 2005, **43**(5), 2291–2302.
144. J. C. Caoili, A. Mayorova, D. Sikes, L. Hickman, B. B. Plikaytis and T. M. Shinnick, Evaluation of the TB-Biochip oligonucleotide microarray system for rapid detection of rifampin resistance in Mycobacterium tuberculosis, *J. Clin. Microbiol.*, 2006, **44**(7), 2378–2381.
145. M. M. Wade, D. Volokhov, M. Peredelchuk, V. Chizhikov and Y. Zhang, Accurate mapping of mutations of pyrazinamide-resistant Mycobacterium tuberculosis strains with a scanning-frame oligonucleotide microarray, *Diagn. Microbiol. Infect. Dis.*, 2004, **49**(2), 89–97.
146. K. J. Cheung, V. Badarinarayana, D. W. Selinger, D. Janse and G. M. Church, A microarray-based antibiotic screen identifies a regulatory role for supercoiling in the osmotic stress response of Escherichia coli, *Genome Res.*, 2003, **13**(2), 206–215.
147. E. Mongodin, J. Finan, M. W. Climo, A. Rosato, S. Gill and G. L. Archer, Microarray transcription analysis of clinical Staphylococcus aureus isolates resistant to vancomycin, *J. Bacteriol.*, 2003, **185**(15), 4638–4643.
148. J. Qiu, D. Zhou, L. Qin, Y. Han, X. Wang, Z. Du, Y. Song and R. Yang, Microarray expression profiling of Yersinia pestis in response to chloramphenicol, *FEMS Microbiol. Lett.*, 2006, **263**(1), 26–31.
149. J. Qiu, D. Zhou, Y. Han, L. Zhang, Z. Tong, Y. Song, E. Dai, B. Li, J. Wang, Z. Guo, J. Zhai, Z. Du, X. Wang and R. Yang, Global gene

expression profile of Yersinia pestis induced by streptomycin, *FEMS Microbiol. Lett.*, 2005, **243**(2), 489–496.

150. P. D. Rogers and K. S. Barker, Evaluation of differential gene expression in fluconazole-susceptible and -resistant isolates of Candida albicans by cDNA microarray analysis, *Antimicrob. Agents Chemother.*, 2002, **46**(11), 3412–3417.

151. S. Kastner, V. Perreten, H. Bleuler, G. Hugenschmidt, C. Lacroix and L. Meile, Antibiotic susceptibility patterns and resistance genes of starter cultures and probiotic bacteria used in food, *Syst. Appl. Microbiol.*, 2006, **29**(2), 145–155.

152. M. Ma, H. Wang, Y. Yu, D. Zhang and S. Liu, Detection of antimicrobial resistance genes of pathogenic Salmonella from swine with DNA microarray, *J. Vet. Diagn. Invest.*, 2007, **19**(2), 161–167.

CHAPTER 7
Whole-cell Sensing Systems in Chemical and Biological Surveillance

ELISA MICHELINI, LUCA CEVENINI,
LAURA MEZZANOTTE AND ALDO RODA

Department of Pharmaceutical Sciences, University of Bologna, Via
Belmeloro 6, 40126 Bologna, Italy

7.1   Introduction

With the increasing threat of biological and chemical warfare agents, developing innovative strategies for rapid, simple and precise detection of these harmful agents is critically important. In response to this demand, much effort has been focused on developing biosensors and biomimetic systems that are well suited for toxicity monitoring of both known and unknown analytes. In particular, the use of living cells (*e.g.*, layers of confluent cells, networks and arrays of living cells) as the sensor elements in such systems has undergone considerable developments during the last decade being used for proof-of-principle studies or being incorporated into prototype devices.

Cell-based screening approaches are gaining more and more diffusion due to the possibility to perform a wide number of different functional cellular assays and the availability of automated screening platforms and bioinformatics tools. Cell-based detection systems found important applications in the fields of hygiene, public safety and security including fighting bioterrorism, for the *in-situ* detection of chemical and biological contaminants, *e.g.*, microorganisms, spores and viruses.[1]

Nano and Microsensors for Chemical and Biological Terrorism Surveillance
Edited by Jeffrey B.-H. Tok
© Royal Society of Chemistry, 2008
Published by the Royal Society of Chemistry, www.rsc.org

7.2 Cell- and Tissue-based Detection Systems

Cell- or tissue-based detection systems exploit the intrinsic ability of a specific cell type to respond to a potentially toxic or infectious agent. In these devices the sensing system, *i.e.* the cell, produces a signal that can be measured by an electrode or optical detector.[2] The cells may derive from a unicellular organism or a specific tissue type such as neurological and cardiac tissues.[3] One way of acquiring cellular functional information for biosensor applications involves extracellular recording from excitable cells, which can generate non-invasive and long-term measurements useful for the detection of harmful substances.

In particular the ability to detect compounds able to affect neurobehavior is of crucial importance to both civilian and military communities and cultured neuronal networks proved to be very effective systems for the monitoring of known and unknown threat agents. For example a neuronal network biosensor based on cultured mammalian neurons grown over microelectrode arrays was developed for the detection of marine toxins.[4] Spinal cord neuronal networks were isolated from embryonic mice and the mean spike rate across the network was analyzed before and during exposure to the toxins. Extracellular action potentials from the network were highly sensitive not only to purified saxitoxin (STX) and brevetoxin (PbTx-3), but also when in combination with complex matrixes such as natural seawater and algal growth medium. By monitoring extracellular action potentials, detection limits of 0.33 and 0.031 nM were obtained for STX and PbTx-3, respectively.

A portable system, incorporating neuronal networks cultured on microelectrode arrays (MEAs), tailored to monitor neuronal extracellular potentials was developed by Pancrazio *et al.*[5] To assess the analytical performance and potential applicability of this system well-known ion channel blockers, tetrodotoxin and tityustoxin, were used. A limitation of this portable MEA is that the fluidics system does not support more than one neuronal network, whereas a dual neuronal network format would be extremely useful as an internal control for environmental fluctuations that could affect the system.

A whole-cell-based biosensor was also used for the detection of *Staphylococcal* alpha toxin using a confluent monolayer of human umbilical vein endothelial cells seeded onto the surface of an asymmetric cellulose triacetate membrane of an ion-selective electrode. This sensor takes advantage of endothelial cell permeability dysfunction to detect the presence of small quantities of permeability-modifying agents such as *S. aureus* alpha toxin with a limit of detection of as low as 0.1 ng/ml after only 20 min of exposure time.[6]

An elegant label-free approach was undertaken by Notingher *et al.*, who developed a Raman spectroscopy cell-based biosensor for rapid detection of toxic agents.[7] This technology allows the monitoring of the biochemical properties of living cells over long periods of time by measuring the Raman spectra of the cells non-invasively, rapidly and without use of labels (Figure 7.1). The Raman spectrum of a cell represents an information-rich "fingerprint" of the overall biochemical composition of the cell; thus different toxic agents that initiate different cellular responses and biochemical changes should produce

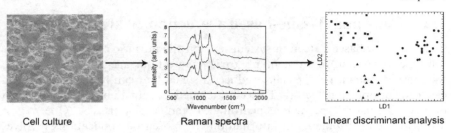

Cell culture   Raman spectra   Linear discriminant analysis

Figure 7.1 Raman Spectroscopy cell-based biosensor for label-free detection of toxic agents.

distinct changes in the Raman spectra. By using multivariate statistical methods, such as principal component analysis (PCA) and linear discriminant analysis (LDA), for analyzing Raman spectra of a human alveolar epithelial cell line, Notingher *et al.* discriminated between the cellular effects of ricin and sulfur mustard, two toxic agents of bioterrorism and chemical warfare significance. The PCA–LDA analysis showed that damaged cells can be detected with high sensitivity (98.9%) and high specificity (87.7%). Moreover, this method showed high accuracy in identifying the nature of the toxic agent, as 88.6% of the cells treated with sulfur mustard and 71.4% of the cells exposed to ricin were classified correctly.

Another cell type that has been extensively investigated for biological and chemical surveillance is chromatophores, neuron-like cells containing pigment granules that are responsible for the brilliant colors of fish, amphibians, reptiles and cephalopods.[8,9] Fish chromatophores from *Betta splendens* were used as the cytosensor element in the development of a portable microscale device capable of detecting certain environmental toxins and bacterial pathogens by monitoring changes in pigment granule distribution. This biosensor incorporating chromatophores was able to detect the presence of certain polynuclear aromatic hydrocarbons at concentrations lower than the Environment Protection Agency (EPA) 550.1 standards. A miniaturized culture chamber was also specifically designed to support chromatophore viability for as long as 3 months.

Tissue-based biosensors obtained from immobilized photosynthetic microorganisms have also been developed for the detection of airborne chemical warfare agents.[10,11] This type of biosensor relies on the fluorescence induction by living photosynthetic tissue. Fluorescence induction curves from photosynthetic organisms such as a unicellular green alga, *Chlorella vulgaris*, and a cyanobacterium, *Nostoc commune*, were reported to be good indicators of the presence of environmental stresses surrounding the organisms and the effects those stresses have on the photosynthetic apparatus. The fluorescence detection system compared fluorescence emissions from algae and cyanobacteria exposed to a "clean" stream of air with air carrying a toxic chemical agent and used this information to calculate the efficiency of Photosystem II photochemistry.

Different chemical warfare agents were tested with this biosensor and changes in total fluorescence yields were evident with the nerve agent Tabun.

These biosensors may be used as continuous rapid-warning sentinels for detection of chemical warfare agents with potential integration into commercially available hand-held fluorescence instrumentation for field applications.

These cell-based biosensors are not as specific as other biosensors based on antibody recognition or nucleic acid hybridization but this feature could be advantageous when the threat agent is unknown.

7.3   Genetically Engineered Whole-cell Sensing Systems

Cell-based assays provide an effective tool for detection of chemical and biological warfare agents, especially in situations where the compounds are part of a complex mixture or in different forms such as natural or synthetic derivatives or bioactive metabolites. Advances in the engineering of functional responses in cells provide a means to refine the response to given agents.

Genetically engineered cells (bacteria, yeasts or mammalian cells) able to produce a signal (*e.g.* fluorescent, bioluminescent, *etc.*) in response to a target analyte represent powerful analytical tools for the routine monitoring of the environment and food for biological and chemical warfare agents, being characterized by low cost and high rapidity and sensitivity.[12] The cells are modified by introducing a reporter gene fused to a regulatory DNA sequence that is activated only in the presence of the analyte of interest, which thus regulates the reporter gene expression (Figure 7.2). The ability to measure the bioavailable fraction of a compound (*i.e.* the fraction of compound able to enter live cells and activate the specific response pathways) is a peculiar feature

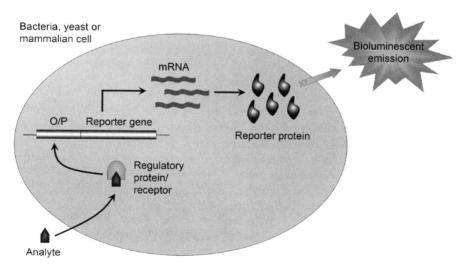

Figure 7.2   Reporter gene technology applied to the development of a bioluminescent whole-cell biosensor.

of whole-cell biosensors, thus providing crucial information that is difficult to obtain with other methodologies.

Several cell-based assays were developed by genetically engineering bacteria to express organophosphorus hydrolase (OPH) on the cell surface.[13]

A biosensor was developed by immobilizing on the carbon paste electrode and genetically engineering a p-nitrophenol degrader, *Pseudomonas putida* JS444, to express OPH.[14] Surface-expressed OPH catalyzed hydrolysis of the p-nitrophenyl substituent organophosphorus pesticides such as paraoxon, parathion and methyl parathion to release p-nitrophenol, which was subsequently degraded by the enzymatic machinery of *P. putida* JS444. The electrooxidization current of the intermediates was measured and correlated to the concentration of organophosphates. The biosensor measured as low as 0.28 ppb of paraoxon, 0.26 ppb of methyl parathion and 0.29 ppb parathion. These detection limits are comparable to cholinesterase inhibition-based biosensors. Unlike the inhibition-based format, this biosensor manifests a selective response to organophosphate pesticides with a p-nitrophenyl substituent only, has a simplified single-step protocol with short response time and can be used for repetitive/multiple and on-line analysis.

More recent approaches involve the use of reporter genes with optical detection (bioluminescent or fluorescent). Among the advantages of cell-based biosensors that employ such reporter genes is signal amplification due both to multiple mRNA copies being transcribed from each reporter gene copy and to multiple reporter protein copies being synthesized from each mRNA molecule. In particular, the green fluorescent protein (GFP), a highly fluorescent protein originally isolated from the jellyfish *Aequorea victoria*, presents various advantages: low toxicity to host cells, direct detection without addition of substrates and the possibility to be expressed in bacteria where it spontaneously folds in its active conformation. Among bioluminescent (BL) reporter genes, luciferase from the North American firefly *Photinus pyralis* is by far the most employed. Luciferase does not require any post-translational modification for enzyme activity and it is not toxic even at high concentrations, being thus suitable for *in vivo* applications in prokaryotic and eukaryotic cells.

Fairey and colleagues developed a reporter gene assay as a direct evolution of previously reported cytotoxicity assays for algal-derived toxins. *c-fos* was selected as biomarker for localizing the effects of toxins for its ability to be induced in neurons of mammals and fish as a result of neuronal stimulation. A mouse neuroblastoma cell line was stably transfected with a *c-fos*-luciferase reporter vector and brevetoxin-1 caused a concentration-dependent increase in luciferase activity with a half-maximal effect that occurred at a concentration comparable to that obtained by direct cytotoxicity assays.[15] More recently the same authors reported a modification of the assay by using human embryonic kidney cells (HEK-293) stably transfected with a human heart voltage-dependent Na($+$) channel instead of mouse neuroblastoma cells.[16]

Besides specific threats, one of the major concerns in case of a biological or chemical attack is the vulnerabilities of water supplies to intentional contamination.[17] The development of rapid systems able to detect the presence of toxic compounds in water samples would be of great relevance. Diverse whole-cell biosensors were developed for the detection of environmental pollution and toxicity.[18] These biosensors are constructed through the fusion of promoters, responsive to the relevant environmental conditions, to easily monitored reporter genes.[19]

Another BL reporter gene, the bacterial luciferase (lux) was fused to different stress-responsive promoters to engineer *Escherichia coli* strains in order to develop a panel of whole-cell biosensors to assess the potential toxicity of water samples.[20]

A sensor that uses engineered B lymphocytes that emit light within seconds of exposure to specific bacteria and viruses was developed by Rider *et al.*[21] B cell lines were engineered to express cytosolic aequorin, a calcium-sensitive biolu-minescent protein from the *Aequorea victoria* jellyfish, as well as membrane-bound antibodies specific for pathogens of interest.[22] Cross-linking of the antibodies by even low levels of the appropriate pathogen elevated intracellular calcium concentrations within seconds, causing the aequorin to emit light. A feature of this biosensor is that the antibody expressed determines the cell specificity and can be tailored to a desired application, although it suffers antibody cross-reactivity problems as other antibody-based technologies (Figure 7.3).

The unlimited potential of the olfactory receptors (ORs) was also explored to detect innumerable chemical agents with great sensitivity and selectivity. The exquisite sensitivity of the olfactory signaling system is presumably due to the presence of multiple ORs responding to a single ligand.[23] Unfortunately, due to inefficient receptor insertion into the plasma membrane, the expression

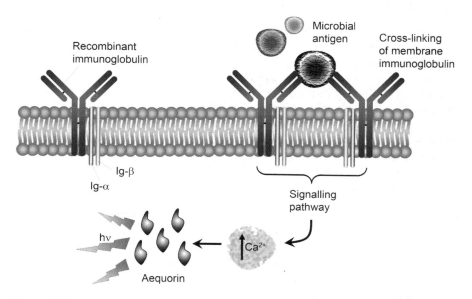

Figure 7.3   Schematic view of a B-cell based sensor for rapid identification of pathogens.

of ORs in heterologous cell systems has been almost unsuccessful so far. Only a handful of mammalian ORs have been functionally expressed in heterologous systems. Functional expression of the rat I7 OR and its trafficking to the plasma membrane was first achieved in the yeast *Saccharomyces cerevisiae* and demonstrated by confocal immunofluorescence microscopy by Minic *et al.*[24]

A novel biosensor for odorant screening using luciferase as a functional reporter was therefore developed and probed with an array of odorants to demonstrate the specificity and selectivity towards ligands. This rapid and inexpensive screening assay was characterized by an extended dynamic range, with the potential of investigating many orphan ORs against the extra-ordinarily large number of natural and synthetic odorants.

More recently, an *S. cerevisiae* strain was engineered to couple the mammalian olfactory receptor signaling to green fluorescent protein expression. By expressing a library of ligand-binding pockets of "orphan" receptors using a receptor scaffold, the identity of the receptor specific for a particular ligand could potentially be defined. Therefore, this strategy was used to identify an OR for a specific odorant: 2,4-dinitrotoluene (DNT), a mimic for the explosive trinitrotoluene (TNT). A library of cDNA inserts was cloned encoding the ligand-binding domains of rat ORs (derived from rat olfactory epithelium) and transfected them into yeast cells. By exposing these cells to DNT ($50\,\mu M$) and scoring the cells that emitted green fluorescence, several DNT-responsive clones were identified. In the near future, this "olfactory yeast" could be useful to detect TNT and similar toxic agents in the environment screened.[25]

7.4   Phage Display Technology

The development of systems for the routine monitoring of the environment and food for biological threat agents is a continuous challenge since the number of potential threat agents is almost infinite.

New techniques for generating diagnostic probes which may meet the strong criteria for biological monitoring are nowadays available. They include combinatorial chemistry, phage display and directed molecular evolution.[26]

The phage display technology is based on the concept that a foreign coding sequence can be spliced in-frame into a phage coat protein gene, so that the "guest" peptide encoded by that sequence is fused to a coat protein, and thus displayed on the exposed surface of the virion. Phage display libraries can be created being an ensemble of up to about 10 billion such phage clones, each harboring a different foreign coding sequence, and therefore displaying a different guest peptide on the virion surface. The foreign coding sequence can be of natural or synthetic origin. Then, a target binding molecule is immobilized on a solid support and exposed to the phage display library. Only phage particles whose displayed peptides bind the immobilized target are captured on the support and can remain there while all other phages are washed away. The captured phage is then eluted from the support and propagated or cloned by infecting bacterial host cells. After several rounds of affinity selection, individual phage clones are propagated and their ability to bind the selector confirmed. By using phage display technology, it is possible to originate phage antibodies, a special type of phage-display construct in which the displayed peptide is an antibody molecule.[27]

There are several examples where phage antibodies have been successfully used in different detection platforms.[28] Once antibodies are selected from phage-display libraries they can be genetically fused to marker proteins (*e.g.*, horseradish peroxidase or alkaline phosphatase) and be used for one-step immunodetection of biological agents. A phage-antibody library using mRNA from mice immunized with botulinum toxin was constructed by Emanuel *et al.*[29] The selected antibody showed better performance than monoclonal antibody in a variety of assay formats including surface plasmon resonance and flow cytometry.

Phage display technology has also been used to develop alternative therapeutic approaches for biothreat agents such as passive immunization, which possesses several important advantages over active vaccination and the use of antibiotics, as it can provide immediate protection against several pathogens such as *Bacillus anthracis*. The selection and characterization of several human monoclonal neutralizing antibodies against the toxin of *B. anthracis* from a phage displayed human scFv library was recently reported.[30] A total of 15 clones were selected with distinct sequences and high specificity to protective antigen and thus were the subject of a series of both biophysical and cell-based cytotoxicity assays. From this panel of antibodies a set of neutralizing antibodies were identified, of which clone A8 recognizes the lethal (and/or edema) factor binding domain, and clones F1, G11 and G12 recognize the cellular receptor binding domain found within the protective antigen).

7.5 New Perspectives in Whole-cell Biosensors for Detecting Threat Agents

Biosensors and particularly whole-cell biosensors have become an important tool in the biosecurity and military sectors.[31,32] The high number of published articles and patents that have been issued and/or are pending suggests that whole-cell biosensors will play a growing role in the near future.[33] To solve most of the problems related to such bioassays, which are commonly time-consuming and require the maintenance of cell cultures,[34] new strategies are now emerging.

An important trend is the development of miniaturized devices based on microfluidics systems and nanotechnologies.[35] Miniaturization of biosensing systems can enhance their utility by decreasing reagent consumption and analysis time and by allowing for the high-throughput screening of samples. Besides reducing the production and development costs, miniaturization consents to develop portable systems for *in-situ* detection, thus meeting the "detect-to-warn" needs for first responders such as soldiers and medical personnel. Portable biosensors for on-site monitoring have been developed using immobilized cells, freeze-dried biosensing strains or cell networks for high-throughput analysis.[5,36–38] Moreover, in the future, the use of single-cell biosensors will allow detailed analyses of samples. Signals from such sensors could be detected with digital imaging, epifluorescence microscopy and/or flow cytometry.

In addition, a new generation of electronic noses for detection and discrimination of volatile compounds, particularly amenable to micro- and nanosensor formats, has been recently envisaged. An olfactory receptor and an appropriate G protein were co-expressed in *S. cerevisiae* cells from which membrane nanosomes were prepared, and immobilized on a sensor chip.[39] The OR stimulation by an odorant was quantitatively evaluated by Surface Plasmon Resonance (SPR), demonstrating that receptor activity was not hampered by immobilization of nanosomes, since selectivity and high sensitivity of a mammalian OR were retained in the device. This chip assay gains benefit from using yeast cells as a host for OR expression (*e.g.*, low cost, simplicity for genetic manipulations) and in addition enables direct and label-free detection of a functional response in operating conditions devoid of living cells. This SPR method is thus suitable for high throughput screening of ORs because immobilized receptors can be stimulated repetitively and an automatic SPR analysis may be introduced.

Together with the development of miniaturized portable systems, the High Content Screening (HCS) approach has a great potential in detecting threat agents. The HCS concept enables simultaneous measurement of multiple features of cellular phenotype that are relevant to therapeutic and toxic activities of compounds.[40–42] Diverse commercial image-based cell screening platforms comprising fluorescent reagents, automated image acquisition hardware, image analysis algorithms and informatics tools are already available with great potential in biological and chemical surveillance.

Acronyms

MEAs	Micro-Electrode Assays
PCA	Principal Component Analysis
LDA	Linear Discriminant Analysis
OPH	Organophosphorus Hydroxylase
GFP	Green Fluorescent Protein
BL	Bioluminescent
ORs	Olfactory Receptors
DNT	2,4-dinitrotoluene
TNT	Trinitrotoluene
SPR	Surface Plasmon Resonance
HCS	High Content Screening

References

1. Y. Bhattacharjee, *Science*, 2005, **309**, 1810.
2. G. T. A. Kovacs, *Proceedings of the IEEE*, 2003, **91**, 915.

3. S. A. Gray, J. K. Kusel, K. M. Shaffer, Y. S. Shubin, D. A. Stenger and J. J. Pancrazio, *Biosens. Bioelectron.*, 2001, **16**, 535.
4. N. V. Kulagina, C. M. Mikulski, S. Gray, W. Ma, G. J. Doucette, J. S. Ramsdell and J. J. Pancrazio, *Environ. Sci. Technol.*, 2006, **40**, 578.
5. J. J. Pancrazio, S. A. Gray, Y. S. Shubin, N. Kulagina, D. S. Cuttino, K. M. Shaffer, K. Eisemann, A. Curran, B. Zim, G. W. Gross and T. J. O'Shaughnessy, *Biosens. Bioelectron.*, 2003, **18**, 1339.
6. G. Ghosh, L. G. Bachas and K. W. Anderson, *Anal. Bioanal. Chem.*, 2007, **387**, 567.
7. I. Notingher, C. Green, C. Dyer, E. Perkins, N. Hopkins, C. Lindsay and L. L. Hench, *J. R. Soc. Interface*, 2004, **1**, 79.
8. F. W. Chaplen, R. H. Upson, P. N. Mcfadden and W. Kolodziej, *Pigment Cell Res.*, 2002, **15**, 19.
9. V. Sharma, A. Narayanan, T. Rengachari, G. C. Temes, F. Chaplen and U. K. Moon, *Biosens. Bioelectron.*, 2005, **20**, 2218.
10. M. Rodriguez Jr, C. A. Sanders and E. Greenbaum, *Biosens. Bioelectron.*, 2002, **17**, 843.
11. C. A. Sanders, M. Rodriguez Jr and E. Greenbaum, *Biosens. Bioelectron*, 2001, **16**, 439.
12. S. F. D'Souza, *Biosens. Bioelectron.*, 2001, **16**, 337.
13. P. Mulchandani, W. Chen, A. Mulchandani, J. Wang and L. Chen, *Biosens. Bioelectron.*, 2001, **16**, 433.
14. Y. Lei, P. Mulchandani, J. Wang, W. Chen and A. Mulchandani, *Environ. Sci. Technol.*, 2005, **39**, 8853.
15. E. R. Fairey, J. S. Edmunds and J. S. Ramsdell, *Anal. Biochem.*, 1997, **251**, 129.
16. E. R. Fairey, M. Y. Bottein Dechraoui, M. F. Sheets and J. S. Ramsdell, *Biosens. Bioelectron.*, 2001, **16**, 579.
17. J. B. Nuzzo, *Biosecur. Bioterror.*, 2006, **4**, 147.
18. E. Vetrova, E. Esimbekova, N. Remmel, S. Kotova, N. Beloskov, V. Kratasyuk and I. Gitelson, *Luminescence*, 2007, **22**, 206.
19. E. Z. Ron, *Curr. Opin. Biotechnol.*, 2007, **18**, 252.
20. R. Pedahzur, B. Polyak, R. S. Marks and S. Belkin, *J. Appl. Toxicol.*, 2004, **24**, 343.
21. T. H. Rider, M. S. Petrovick, F. E. Nargi, J. D. Harper, E. D. Schwoebel, R. H. Mathews, D. J. Blanchard, L. T. Bortolin, A. M. Young, J. Chen and M. A. Hollis, *Science*, 2003, **301**, 213.
22. M. J. Cormier, D. C. Prasher, M. Longiaru and R. O. McCann, *Photochem. Photobiol.*, 1989, **49**, 509.
23. P. Mombaerts, *Science*, 1999, **286**, 707.
24. J. Minic, M. A. Persuy, E. Godel, J. Aioun, I. Connerton, R. Salesse and E. Pajot-Augy, *FEBS J.*, 2005, **272**, 524.
25. V. Radhika, T. Proikas-Cezanne, M. Jayaraman, D. Onesime, J. H. Ha and D. N. Dhanasekaran, *Nat. Chem. Biol.*, 2007, **3**, 325.
26. V. A. Petrenko and I. B. Sorokulova, *J. Microbiol. Methods*, 2004, **58**, 147.
27. V. A. Petrenko and V. J. Vodyanoy, *J. Microbiol. Methods.*, 2003, **53**, 253.

28. Y. Fujinami, Y. Hirai, I. Sakai, M. Yoshino and J. Yasuda, *Microbiol. Immunol.*, 2007, **51**, 163.
29. P. Emanuel, T. Obrien, J. Burans, B.R. Dasgupta, J.J. Valdes and M. Eldefrawi, *J. Immunol. Methods*, **193**, 189.
30. B. Zhou, C. Carney and K. D. Janda, *Bioorg. Med. Chem.*, 2008, **16**, 1903.
31. B. Pejcic, R. De Marco and G. Parkinson, *Analyst*, 2006, **131**, 1079.
32. D. V. Lim, J. M. Simpson, E. A. Kearns and M. F. Kramer, *Clin. Microbiol. Rev.*, 2005, **18**, 583.
33. O. A. Sadik, A. K. Wanekaya and S. Andreescu, *J. Environ. Monit.*, 2004, **6**, 513.
34. S. Cai, B. R. Singh and S. Sharma, *Crit. Rev. Microbiol.*, 2007, **33**, 109.
35. S. E. Ong, S. Zhang, H. Du and Y. Fu, *Front. Biosci.*, 2008, **13**, 2757.
36. F. Heer, S. Hafizovic, T. Ugniwenko, U. Frey, W. Franks, E. Perriard, J. C. Perriard, A. Blau, C. Ziegler and A. Hierlemann, *Biosens. Bioelectron.*, 2007, **22**, 2546.
37. T. J. O'Shaughnessy, S. A. Gray and J. J. Pancrazio, *J. Appl. Toxicol.*, 2004, **24**, 379.
38. K. H. Gilchrist, V. N. Barker, L. E. Fletcher, B. D. DeBusschere, P. Ghanouni, L. Giovangrandi and G. T. Kovacs, *Biosens. Bioelectron.*, 2001, **16**, 557.
39. J. M. Vidic, J. Grosclaude, M. A. Persuy, J. Aioun, R. Salesse and E. Pajot-Augy, *Lab Chip*, 2006, **6**, 1026.
40. S. B. Tencza and M. A. Sipe, *J. Appl. Toxicol.*, 2004, **24**, 371.
41. D. W. Young, A. Bender, J. Hoyt, E. McWhinnie, G. W. Chirn, C. Y. Tao, J. A. Tallarico, M. Labow, J. L. Jenkins, T. J. Mitchison and Y. Feng, *Nat. Chem. Biol.*, 2007, **4**, 59.
42. N. Ye, J. Qin, W. Shi, X. Liu and B. Lin, *Lab Chip*, 2007, **7**, 1696.

CHAPTER 8

Conducting Polymer Transistors for Sensor Applications

FABIO CICOIRA,[a] DANIEL A. BERNARDS[b] AND
GEORGE G. MALLIARAS[b]

[a] Department of Materials Science and Engineering, Bard Hall, Cornell
University, Ithaca 14850, USA; Also at IFN-CNR, via alla Cascata 56/c,
38050 Povo (Trento), Italy; [b] Department of Materials Science and
Engineering, Bard Hall, Cornell University, Ithaca 14850, USA

8.1 Introduction

Organic semiconductors have attracted enormous attention during the last
decades due to their unique properties, such as ease of processing and tunability
of electronic properties through chemical synthesis.[1] The three most well-
known classes of devices based on organic semiconductors are organic light
emitting diodes (OLEDs), organic thin film transistors (OTFTs) and organic
solar cells. Enormous progress has been made in the field of organic semiconductors, as demonstrated by the recent commercialization of displays based
on OLEDs. Chemical and biological sensing is another promising application
of organic semiconductors and, although there has been some low-level activity
in this field for a while, it is only in the past few years that the interest is
becoming significant. We predict that in the near future organic-based sensors
will become a main thrust of organic electronics and play a pivotal role in the
emerging field of organic bioelectronics.[2]

A sensor is a device that is composed of a recognition element, which
interacts selectively with a biological or chemical species, and a transducer,
which translates this interaction into an observable signal. Examples of

Nano and Microsensors for Chemical and Biological Terrorism Surveillance
Edited by Jeffrey B.-H. Tok
© Royal Society of Chemistry, 2008
Published by the Royal Society of Chemistry, www.rsc.org

recognition elements are antibodies, bacteriophages and oligonucleotides. There are several different types of transducers, including electrochemical, optical, piezoelectric and thermal.

Recent demonstrations of sensor concepts using OFETs[3] and OLEDs[4] paved the way for developments in this field. OTFTs in particular are excellent candidates for transducers in sensor applications due to their simple electrical readout, inherent signal amplification, straightforward miniaturization and facile incorporation into arrays and circuits. OTFTs are currently being explored in sensors for mechanical deformation, pressure, moisture and organic vapors, pH and ion concentrations, as well as a variety of biological analytes.[5]

Sensors are of high demand in our society, and there is an urgent need for fast, highly sensitive, portable and inexpensive sensors for a broad range of applications. The latter include medical diagnostics, water and food safety, detection of chemical and biological warfare agents and environmental monitoring. An example where sensors have made a large impact is in diabetes management. Easy to use and relatively inexpensive devices that measure glucose concentration are now widely available. The hope is that novel sensor technologies will help extend the application of such diagnostics to the detection of multiple pathogens and disease markers as well as make them less invasive.

In this chapter we will focus on a particular type of OTFT called the organic electrochemical transistor (OECT). We wish to stress that this is not meant to be a literature review. Rather, we are following a didactic approach where we discuss the mechanism of operation of OECTs and its connection to sensor performance.

8.2   The Organic Electrochemical Transistor (or Conducting Polymer Transistor)

8.2.1   History

The first OECT was reported in the 1980s by Wrighton and co-workers and employed polypyrrole as the active material.[6] Three Au microelectrodes covered with polypyrrole were immersed in an electrolyte solution of 0.1 M n-Bu$_4$NClO$_4$ in acetonitrile (inset of Figure 8.1). The outer Au electrodes were used to measure the current flow through organic film, and a gate potential was applied between the central Au electrode and a counter electrode immersed in the electrolyte.

The output characteristics of a polypyrrole ECT are shown in Figure 8.1. At a negative gate potential (V_g), where the polypyrrole is in its insulating state, no drain current was detected (the transistor was in the *off state*). When the gate potential was increased, the polypyrrole was oxidized. As a result, the device turned on, and a significant drain current flowed through the polypyrrole film. This work paved the way for a series of OECTs based on different conducting polymers (*e.g.* polyaniline,[7-11] polypyrrole,[12,13] polycarbazole,[14] polythiophene and their derivatives[15,16]), with potential applications covering a broad spectrum of chemical and biological sensing.

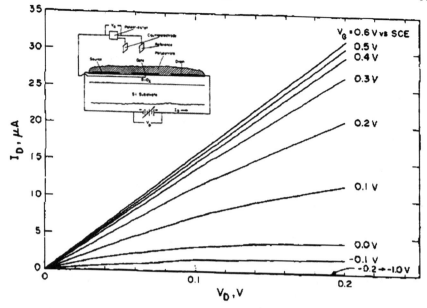

Figure 8.1 Output characteristics of a polypyrrole OECT measured in $CH_3CN/$ 0.1 M n-Bu$_4$NClO$_4$. Inset: cross-sectional view of the device and representation of the circuit elements used to characterize it. The source, drain and gate are 3-μm wide, 140-μm long and 0.12-μm thick Au electrodes coated with about 10^{-7} mol/cm^2 of polypyrrole. Reprinted with permission from [6]. Copyright American Chemical Society.

However, most of the polymers mentioned above may not be suitable for biosensing applications. For instance, polyaniline is stable only within a narrow range of pH, and the conductivity of polypyrrole irreversibly deteriorates in the presence of H_2O_2, a species commonly involved in glucose detection. These limitations have been overcome by the use of poly(3,4-ethylenedioxythiophene) (PEDOT), which has emerged as the most successful conducting polymer for sensing applications.[17] PEDOT is electrochemically active, exhibits high environmental stability and is stable over a broad pH range. Electrical conductivities of 1–100 S/cm are regularly obtained by doping PEDOT (*p-type* doping). The most frequently used counter ion is poly(styrene sulfonate) (PSS). Besides increased conductivity, the addition of PSS allows for a stable suspension of the insoluble PEDOT in water. Aqueous PEDOT:PSS solution (*e.g.* BAYTRON P) is a commercially available, highly p-doped organic semiconductor (structure in Figure 8.2) that can be used for solution deposition of conducting films. Further enhancement of conductivity can be achieved adding to the PEDOT:PSS solution organic compounds such as ethylene glycol, dodecylbenzenesulfonic acid (DBSA), dimethyl sulfoxide (DMSO) or N-N-dimethylformamide (DMF).[18–20] These compounds are thought to induce a screening effect between the positively charged PEDOT chains and the negatively charged PSS chains hence

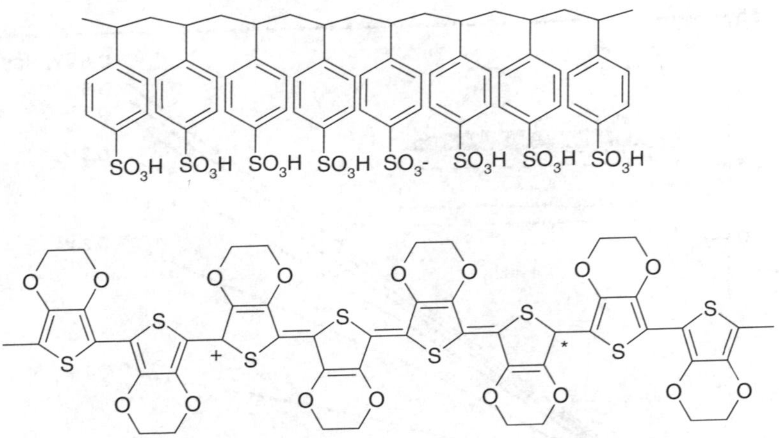

Figure 8.2 Chemical formula of PEDOT:PSS.

reducing the Coulomb interactions between them,[19] as well as to change the PEDOT:PSS film morphology.

In the last decade, Magnus Berggren and colleagues at Linköping University have demonstrated a variety of devices based on PEDOT:PSS ECTs, including logic circuits,[21] bistable transistors,[22] wettability switches,[23] humidity sensors[24] and electronic ion pumps.[25] In these devices, electrodes and active layers are fabricated entirely from PEDOT:PSS, which allows fabrication by high-throughput screen-printing processes on a wide range of substrates (including plastic and paper).

8.2.2 Sensing Applications of OECTs

In OECT-based sensors the polymer layer acts as a transducer. As we discuss below, OECT-based sensors can operate in the non-Faradaic and the Faradaic regimes, the former regime characterized by the absence of steady-state current flow through the gate circuit. The two regimes offer different opportunities for sensor applications. In non-Faradaic operation the OECT acts as an ion-to-electron converter, whereas in the Faradaic regime a redox reaction alters the potential within the electrolyte and this is detected by measuring the drain current in the organic semiconductor film. This mode of operation, typically used for enzymatic sensing, is referred to as remote voltage sensing.

Sensing applications of OECTs were first demonstrated in the mid-1980s by Wrighton and colleagues.[9] This work was followed by the realization of sensors employing different active layers (mainly PEDOT, polyaniline, polypyrrole, polycarbazole and polythiophenes) as transducers for detection of a wide range of chemical and biological species, such as DNA,[26] IgG antigen-antibody,[15] H_2O (humidity),[24] urea,[27–30] metal ions,[31] H_2O_2,[32,33] O_2,[9] protons (pH sensing),[12] hemoglobin,[27] penicillin[12] and NADH.[13] Recently our group has reported enzyme-based[34] and ion channel-based sensors[35] utilizing PEDOT:PSS ECTs as well as the integration of these devices in microfluidic systems.[36]

8.2.3 Mechanism of Operation

A quantitative understanding of the operation mechanism of OECTs is tantamount to optimizing the response of OECT-based sensors. A simple model of operation for *p-type* OECTs working in non-Faradaic regime has recently been formulated by Bernards and Malliaras.[37] The schematic of such an OECT (*e.g.* an OECT based on PEDOT:PSS) is shown in Figure 8.3. The essential components are the transistor channel, the electrodes (source, drain and gate) and an electrolyte in contact with the channel and the gate. The channel is typically a thin film of a semiconducting polymer in its doped (conducting) state. For this reason OECTs are also called conducting polymer transistors. If a highly conducting polymer is employed, the channel and the electrodes can be made by the same material, which considerably simplifies device processing. The electrolyte medium can be a liquid, a gel or a solid.

Typical electrical characteristics for a PEDOT:PSS OECT are shown in Figure 8.4 (device details are given in the figure caption). As a convention the source is grounded and a voltage relative to the ground is applied to the drain electrode. The current passing through the channel is monitored as a function of the gate voltage. At zero gate voltage the transistor is in its *on state* and a high current passes through the channel. Upon application of a positive gate voltage, cations from the electrolyte permeate the organic semiconductor to give the following electrochemical reaction:[24]

$$PEDOT^+ : PSS^- + M^+ + e^- \rightarrow PEDOT + M^+ : PSS^-$$

where M^+ is a cation and e^- is an electron from the source or drain electrode (depending on the sign of V_d). The reaction results in de-doping of the channel, which decreases the drain current. At low drain voltages, the dependence of the current on the drain voltage is linear. As the drain voltage is made more negative, the drain current tends towards saturation (3rd quadrant, Figure 8.4).

The above description also clarifies the main difference between OECTs and organic field-effect transistors (OFETs): the operation of OECTs relies on electrochemical doping/de-doping whereas that of OFETS relies on field-effect doping.

For a simple model, the OECT can be divided into an electronic and an ionic circuit. The ionic circuit accounts for transport of ionic charge in the electrolyte and is described as a combination of linear circuit elements (Figure 8.5). The electronic circuit consists of the *p-type* organic semiconductor film that transports holes between source and drain electrodes whose behavior is described by Ohm's law:

$$J(x) = q \cdot \mu \cdot p(x) \cdot \frac{dV(x)}{dx} \qquad (8.1)$$

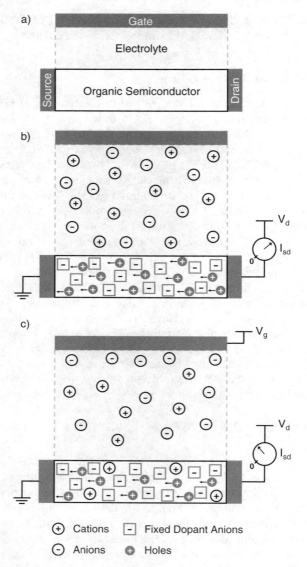

Figure 8.3 Working principle of a PEDOT:PSS OECT. (a) OECT labeled with appropriate naming conventions. (b) OECT without gate voltage applied. (c) OECT with gate voltage (V_g) applied. Current is determined by the extent to which the organic semiconductor is de-doped. Reprinted with permission from [37]. Copyright Wiley VCH.

where J is the current density, q is the elementary charge, μ is the hole mobility, p is the hole density and dV/dx is the electric field. Electronic transport in this circuit depends on hole density and mobility. A de-doping mechanism is used to describe carrier concentrations within the semiconductor upon application of a gate voltage. Cations from the electrolyte permeate the semiconductor film, and each

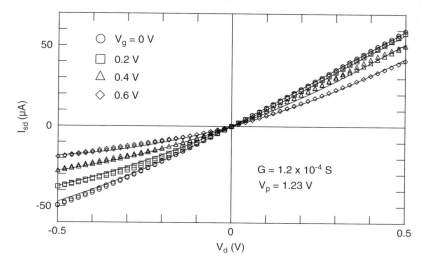

Figure 8.4 Experimental steady-state current *vs.* voltage characteristics (data points) and fit to the model (solid lines) for $G = 1.2 \times 10^{-4}$ S and $V_p = 1.23$ V. A 10-mM NaCl solution was used as the electrolyte. The device had channel length (L) and channel width (W) of 5 mm and 6 mm. Reprinted with permission from [37]. Copyright Wiley VCH.

cation compensates one PSS acceptor. For each compensating cation, a hole extracted at the source is not replaced by injection at the drain (assuming $V_d > 0$).

Using this model, an expression for the effective dopant density in a volume, v, of semiconductor material will be:

$$p = p_0 \cdot \left(1 - \frac{Q}{q \cdot p_0 \cdot v}\right) \tag{8.2}$$

where p_0 is the initial hole density in the organic semiconductor before the application of a gate voltage and Q is the total charge of the cations injected in the organic film from the electrolyte.

Assuming that the gate electrode is ideally polarizable (non-Faradaic regime), the ionic circuit can be described by a resistor (R_s) and a capacitor (C_d) in series.[38] The resistor describes the conductivity of the electrolyte and depends on its ionic strength. The capacitor accounts for polarization at the organic film/electrolyte and gate/electrolyte interfaces. In general the capacitance per unit area of a conducting polymer[39] is significantly greater than that of a Pt gate. As a result, the total capacitance (consider two capacitors in series) will be determined by the gate capacitance. The transient behavior of this element upon the application of a gate voltage exhibits the characteristics of a charging capacitor:

$$Q(t) = Q_{ss} \cdot \left[1 - \exp\left(-\frac{t}{\tau_i}\right)\right] \tag{8.3}$$

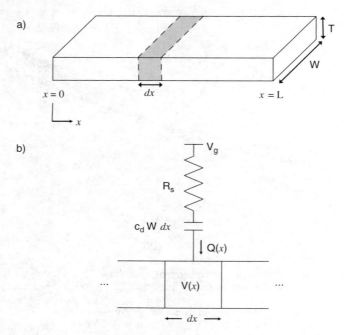

Figure 8.5 Device geometry used in the OECT model. (a) Organic semiconductor film where the source is located at $x=0$ and the drain at $x=L$. (b) Charge (Q) from the ionic circuit is coupled to the voltage in the electronic circuit at a position x along the organic semiconductor. Reprinted with permission from [37]. Copyright Wiley VCH.

where $Q_{ss}=C\,\Delta V$ is the total charge that passes through the circuit, ΔV is the voltage applied across the electrolyte and the ionic transit time is described by $\tau_i=C_d\,R_s$. Because C_d depends on the device area considered, it is convenient to refer to $C_d=c_d\,A$ for much of the analysis, where c_d is capacitance per unit area and A is the area of the device under consideration. For simplicity, the concentration and potential dependence of the ionic double layer capacitance are neglected and a constant value is assumed for c_d.

8.2.4 Steady State in the Non-Faradaic Regime

To solve for OECT device behavior, the effective dopant density (Equation (8.2)) must be spatially known throughout the organic film. If a differential slice, dx, in the vicinity of position x is considered (Figure 8.5), then the charge in that slice at steady state is related to Q_{ss} from Equation (8.3):

$$Q(x) = c_d \cdot W \cdot dx \lfloor V_g - V(x) \rfloor \tag{8.4}$$

where V_g is the gate voltage, $V(x)$ is the spatial voltage profile within the organic film and W is the width of the organic film. Combining Equations (8.1)–(8.4) it is possible to obtain the governing equation for OECT characteristics at steady state:

$$J(x) = q \cdot \mu \cdot p_0 \cdot \left[1 - \frac{V_g - V(x)}{V_p}\right] \cdot \frac{dV(x)}{dx} \qquad (8.5)$$

where V_p is the pinch-off voltage, defined as $q\, p_0\, T/c_{\mathrm{d}}$.

In the first quadrant of Figure 8.4 ($V_d > 0$) there are two regimes of behavior. First, when $V_d < V_g$, de-doping will occur everywhere in the organic film. Using the previous assumptions, Equation (8.5) can be rewritten in terms of current and then solved explicitly, placing the source at $x = 0$ and the drain at $x = L$:

$$I = G \cdot \left[1 - \frac{V_g - \frac{1}{2} \cdot V_d}{V_p}\right] \cdot V_d \qquad (8.6)$$

where G is the conductance of the organic semiconductor film ($G = q\,\mu\,p_0\,W\,T/L$). The second regime occurs when $V_d > V_g$, and de-doping will only occur in the region of the device where $V(x) < V_g$. This regime is described by:

$$I = G \cdot \left[V_d - \frac{V_g^2}{2 \cdot V_p}\right] \qquad (8.7)$$

where the current is linear with drain voltage, and the onset of linear behavior occurs when $V_d = V_g$.

In the third quadrant ($V_d < 0$), it is possible to completely de-dope portions of the organic film when the local density of injected cations becomes equal to the intrinsic dopant density of the semiconducting material. Mathematically this is true when $(V_g - V_d) \geq -V_p$, where the critical drain voltage for saturation can be written as $V_d^{\mathrm{sat}} = V_g - V_p$. Locally the semiconductor will be depleted near the drain contact, but holes injected into this region will still be transported to the drain. An equivalent argument is used to describe saturation in depletion-mode field effect transistors.[40] If the magnitude of V_d increases beyond V_d^{sat}, the extent of the depleted region will move slightly toward the source. For organic films that are sufficiently long, the location of the depleted region nearest the source contact will not change appreciably with V_d and the drain current will saturate. If the extent of the depletion region moves significantly with variation in V_d, the current will not saturate but will continue to increase, an effect that can be observed in devices with short source-drain spacing.[41] In the limit of long channels, for $V_d \leq V_d^{\mathrm{sat}}$, the current will only depend on the drain voltage at saturation for a particular gate voltage:

$$I = -\frac{G \cdot \left(V_d^{sat}\right)^2}{2 \cdot V_p} \tag{8.8}$$

The model yields an excellent fit (solid lines) to experimental steady-state current–voltage characteristics (points) shown in Figure 8.4. Such a fit relies on two parameters. The first is the conductance of the organic semiconductor film ($G = q\mu p_0 W T/L$), which can easily be determined with conventional techniques. The second parameter is the pinch-off voltage ($V_p = q p_0 T/c_d$) and is a measure of the dopant density of the semiconductor film relative to the ionic charge that is leveraged from solution for de-doping. The pinch-off voltage indicates the onset of saturation in the absence of a gate voltage and is akin to the pinch-off voltage in conventional depletion mode field effect transistors.[42]

When OECTs are used for sensing applications it is important to understand the relative, rather than absolute, device response upon gating. Namely, the relevant parameter is $\Delta I_{sd}/I_{sd}$, where ΔI_{sd} is the change in current upon application of a gate voltage. As shown in Figure 8.6, the relative device response is large in the third quadrant of operation and increases with increasing gate voltage. Such characteristics are paramount in developing high-sensitivity sensors and are a useful tool in determining optimal device operating conditions.

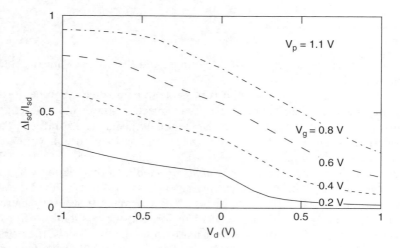

Figure 8.6   Simulated steady-state device response as a function of drain voltage for a series of gate voltages with $V_p = 1.1$ V. I_{sd} is the drain current without an applied gate voltage and ΔI_{sd} is the change in drain current upon the application of a gate voltage. Reprinted with permission from [37]. Copyright Wiley VCH.

8.2.5 Transient Behavior in Non-Faradaic Regime

The transient behavior of OECTs will be dominated by two effects: injection of a cation from the electrolyte into the organic film and removal of a hole at the source electrode ($V_d > 0$). In order to make the calculation of the transient response tractable, the spatial variation of the voltage and the hole density are ignored and an average ionic current and hole density are used. By accounting for the current associated with the removal of holes due to de-doping (in addition to that from Ohm's law), the simplified behavior can be described by:

$$J(t) = q \cdot \mu \cdot p(t) \cdot \frac{V_d}{L} + q \cdot f \cdot L \cdot \frac{dp(t)}{dt} \tag{8.9}$$

where f is a proportionality constant to account for the spatial non-uniformity of the de-doping process. The characteristic range for f is 0 (for instance when $V_d \gg V_g$ at positive V_d) to 1/2 (for instance when $V_g \gg V_d$). Much of the complexity of the time-dependent response is incorporated into f, which is expected to depend on the gate and drain voltages. Using Equation (8.2), the transient response in Equation (8.9) can be determined exactly:

$$I(t) = G \cdot \left(1 - \frac{Q(t)}{q \cdot p_0 \cdot t}\right) \cdot V_d - f \cdot \frac{dQ(t)}{dt} \tag{8.10}$$

where $Q(t)$ is the transient response of the relevant ionic circuit. Under constant gate voltage, the electrolyte model described above (Equation 8.3) is applied. For simplicity, the transient behavior is only described for the case where de-doping occurs everywhere within the organic film without saturation effects. An average voltage drop between the organic film and the gate electrode ($\Delta V = V_g - 1/2V_d$) is chosen to ensure that transient behavior is consistent with steady-state characteristics. Using these assumptions, the transient behavior for a simplified OECT can be described as:[37]

$$I(t, V_g) = I_{ss}(V_g) + \Delta I_{ss} \cdot \left(1 - f \cdot \frac{\tau_e}{\tau_i}\right) \exp\left(\frac{t}{\tau_i}\right) \tag{8.11}$$

where $I_{ss}(V_g)$ is the steady-state drain current at a gate voltage V_g, $\Delta I_{ss} = I_{ss}(V_g = 0) - I_{ss}(V_g)$, $\tau_e = L^2/\mu V_d$ is the electronic transit time and τ_i is the ionic transit time. This gives rise to the transient behavior shown in Figure 8.7: the approach to steady state for an OECT in response to an applied gate voltage can be either a monotonic decay ($\tau_i > f\tau_e$) or a spike-and-recovery ($\tau_I < f\tau_e$).

Qualitatively, a monotonic decay indicates the electronic response of the organic film (*i.e.* how quickly holes can be extracted from the film) is sufficiently fast that it can be ignored when considering the overall transient response. This is typically the case for devices with small source-drain spacing and/or large

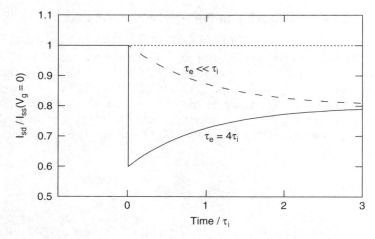

Figure 8.7 Simulated drain current transients for a constant drain voltage with an arbitrary ΔI and fixed geometric factor ($f = 1/2$). The transients demonstrate the two different characteristic responses. Reprinted with permission from [37]. Copyright Wiley VCH.

drain voltages. A spike-and-recovery indicates that hole transport in the organic film occurs at a relatively slow rate and the transient current is dominated by hole extraction from the film. From Equation (8.11), it is apparent that the transient response of an OECT can be characterized primarily by two parameters (τ_i and τ_e) that describe underlying time scales for transport. The characteristic time constant for ionic transport in the electrolyte (τ_i) is determined by the solution resistance and capacitance of the ionic double layer. Using Gouy–Chapman theory for double-layer capacitance (neglecting voltage dependence) along with linear solution conductivity, $\tau_I \sim l/C^{1/2}$ (where l is distance between the organic film and gate electrode and C is the ionic concentration).[38] Qualitatively, decreasing the gate electrode distance from the channel or increasing electrolyte concentrations will lead to improved device response times, which agrees qualitatively with our experimental observations. For example, by varying the applied drain voltage, the character of the transient can be altered as shown in Figure 8.8. As expected, the transient changes from a monotonic decay to a spike-and-recovery with decreasing V_d.

8.3 Operation as Ion-to-Electron Converter

In ion-to-electron converters the devices operate in the transient regime. The presence of the analyte changes the response time of the ionic circuit by changing either the resistance or the capacitance associated with the equivalent circuit of the electrolyte. Two representative examples of ion-to-electron converters are humidity sensors and ion selective membranes, discussed below.

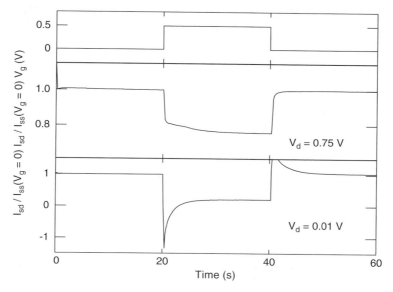

Figure 8.8 Measured drain current transients. Drain current is normalized to the drain current prior to applied gate voltage [$I_{sd}(V_g = 0)$]. Two characteristic responses can be observed with variation in V_d. 10 mM NaCl solution was used as the electrolyte and the organic film dimensions were $L = 5$ mm and $W = 6$ mm. Reprinted with permission from [37]. Copyright Wiley VCH.

8.3.1 Humidity Sensor

Nillson and colleagues reported on an all-organic air humidity sensor based on an OECT.[24] The device employed PEDOT:PSS channel and electrodes in a lateral arrangement with the proton-conductor Nafion as the electrolyte. The PEDOT:PSS pattern was generated by screen-printing on a polyester foil (equivalent results were obtained for devices fabricated on paper). The device showed OECT behavior in an environment where the relative humidity was varied between 25 and 80%. The variation of the ambient humidity level did not result in a considerable variation of the steady-state current but significant variation in the transient characteristics was observed. The sensor response was evaluated in the dynamic mode (*i.e.* measuring the drain current at a given time after application of gate voltage). Figure 8.9 displays the drain current *versus* relative humidity 15 s after application of a gate voltage of 1.2 V. The drain current showed an exponential dependence on the humidity level between 40% and 80% and displayed a lower sensitivity between 25% and 40%.

The sensor response relies on the change of the ion conductivity of Nafion upon exposure to water. The ion conductivity of the Nafion film determined the speed of device response and therefore the extent of current modulation upon application of the gate voltage after a fixed time. Changes in Nafion conductivity also occur upon exposure to other solvents. However, water has a much greater effect, which endows selectivity to humidity.

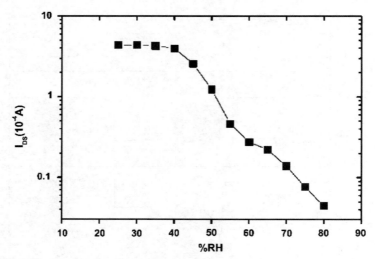

Figure 8.9   Drain current (I_{sd}) measured as a function of relative humidity (RH) after application of a gate voltage of 1.2 V for 15 s. Reprinted with permission from [24]. Copyright Elsevier publishing.

8.3.2   Bilayer Membranes with Ion Channels

Our group reported on the use of bilayer lipid membrane (BLM) with ion channels as the recognition elements in OECTs. The devices employed a PEDOT:PSS channel, Au source and drain electrodes, a Ag/AgCl gate and KCl electrolyte.[35] The BLM was introduced between the gate and the channel of the OECT (Figure 8.10).

The presence of a BLM introduced an additional capacitance in the ionic circuit (see ref. 43) and suppressed gate modulation of I_{sd} during the gate pulse. Modulation upon gating was restored when the applied gate voltage caused rupture of the membrane (for $V_g > 0.3$ V). Alternatively, gating was also restored by the addition of the ion channel gramicidin to the device, where the extent of gating depended on the quantity added (Figure 8.11). Interestingly, in the presence of a $CaCl_2$ electrolyte, the introduction of gramicidin did not re-establish gating. This selectivity is a consequence of the valence-dependent permeability of gramicidin channels, which can be exploited to distinguish between mono- and divalent cations.

8.3.3   Optimization of Device Performance

The performance of ion-to-electron converters discussed above depends on the time (RC) constant of the ionic circuit. To optimize the device performance, it is desirable that the transient response is determined by the ionic circuit (*i.e.* the electronic response of the organic film should be faster than the response of the ionic circuit). This condition is satisfied when $\tau_i > \tau_e$. As discussed above,

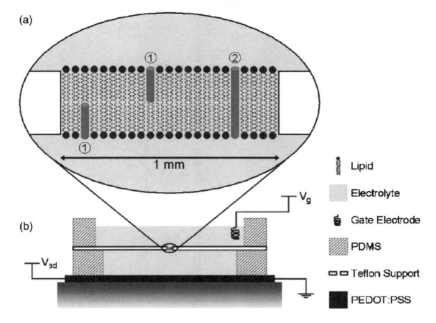

Figure 8.10 Schematic of PEDOT:PSS electrochemical transistor gated through a bilayer lipid membrane (not drawn to scale). (a) Representation of BLM formed on a Teflon support with gramicidin, shown as ion-blocking monomers (1) and an ion permeable dimer (2). (b) Layout for overall device design (some device details omitted for clarity). Reprinted with permission from [35]. Copyright American Institute of Physics.

$\tau_e = L^2/\mu V_d$ and $\tau_i \sim l/C^{1/2}$, which gives $\tau_i/\tau_e \sim l\mu V_d/C^{1/2} L^2$. From this relationship, it is apparent that the device response can be tuned by varying parameters such as gate electrode location, channel length and drain voltage.

8.4 Operation as Remote Voltage Sensor (Enzymatic Sensing)

Enzymatic sensing is used for the detection of glucose in human blood. Highly selective, simple glucose sensors are required for the management of the diabetes mellitus disease. This disease is characterized by variable hyperglycemia, which is caused by low levels or an abnormal reaction to insulin, a blood glucose regulating hormone. As inexpensive and portable glucose monitors are commercially available, further development of glucose sensors does not seem to have much merit. However, glucose detection is the fruit-fly of enzymatic sensing, and new concepts developed for this purpose can be translated to sensing of other metabolites and disease markers for which there is no

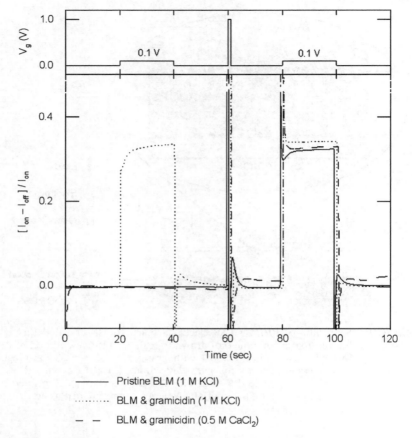

Figure 8.11 Transient response of a PEDOT:PSS electrochemical transistor gated through a BLM with the introduction of gramicidin at $V_d = 0.1$ V. The response with $CaCl_2$ electrolyte in the absence of gramicidin is equivalent to that of a pristine BLM in KCl electrolyte and is omitted for clarity. $I_{on} = I_{sd}$ at $V_g = 0$ V and $I_{off} = I_{sd}$ at $V_g = 0.1$ V. Reprinted with permission from [35]. Copyright American Institute of Physics.

commercially available monitor. Moreover, sensors with a lower detection range might allow non-invasive measurement of glucose.

Our group has demonstrated a simple glucose sensor based on an OECT employing a PEDOT:PSS channel and source/drain electrodes, and a Pt gate electrode immersed in phosphate buffered saline (PBS), as shown in Figure 8.12a.[34] The enzyme glucose oxidase (GOx) was added (free floating) to detect the concentration of glucose present in the electrolyte.

The drain current showed a weak modulation upon application of a gate voltage (at $V_d = 0.2$ V), which remained unchanged after addition of the enzyme GOx (Figure 8.13). Subsequent addition of glucose to the electrolyte caused a dramatic increase in gate modulation, as shown in Figure 8.13. The sensor

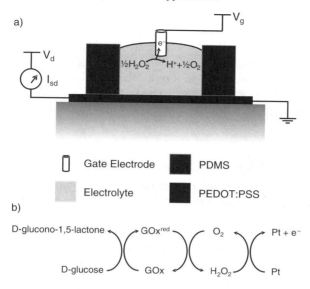

Figure 8.12 (a) Schematic of a typical organic electrochemical transistor (not drawn to scale). The reaction of interest is shown at the gate electrode. (b) Cycle of reactions involved in glucose sensing. Reprinted with permission from [44]. Copyright Royal Society of Chemistry.

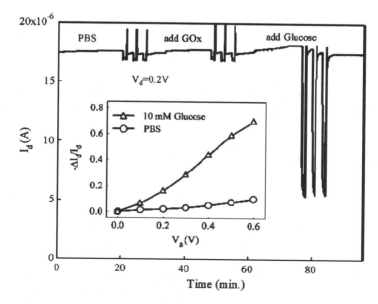

Figure 8.13 I_d *vs.* time for the glucose sensor in PBS solution, in which first GOx and then glucose are added. $V_d = 0.2$ V, and V_g is pulsed to 0.6 V for 1 min. Inset shows the relative change of I_d. Reprinted with permission from [34]. Copyright Royal Society of Chemistry.

response (inset of Figure 8.13) was shown to depend linearly on gate voltage for glucose concentrations between 0.1 and 1 mM.

The sensor response is due to an electrochemical effect and follows the reaction cycle shown in Figure 8.12b. Oxidation of glucose by GOx produces H_2O_2, which can be oxidized to O_2 at the Pt electrode. This reaction is accompanied by de-doping of the PEDOT:PSS channel. To understand how this works one needs to consider device operation in the Faradaic regime.

8.4.1   Model of Operation for Faradaic Regime

Understanding the operational mechanism of OECT-based enzymatic sensors is paramount because of their potential applications in healthcare. Our group has recently published a simple model for OECTs in the Faradaic regime.[44] As discussed above, introduction of glucose to the OECT affects the drain current to an extent depending on gate voltage and glucose concentration. The transfer characteristics of the devices (Figure 8.14a) reveal that the magnitude of the modulation increases with increasing glucose concentration.

Interestingly, the data in Figure 8.14a can be scaled to yield a universal curve (Figure 8.14b), where the gate voltage is scaled to an effective gate voltage according to: $V_g^{eff} = V_g + V_{offset}$. V_{offset} is an offset voltage that has a logarithmic dependence on glucose concentration up to 1 mM, which tapers off at higher concentrations (Figure 8.15). Logarithmic behavior is reminiscent of the Nernst

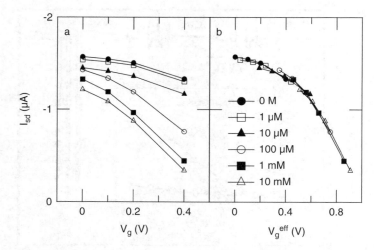

Figure 8.14   (a) Drain current plotted as a function of applied gate voltage for a fixed drain voltage ($V_d = -0.2$ V) and various glucose concentrations. (b) Drain current plotted as a function of effective gate voltage, where the applied gate voltage is shifted by a constant that depends on concentration such that the measured current lies along a universal curve. The extent of the shift is determined by glucose concentration. Reprinted with permission from [44]. Copyright Royal Society of Chemistry.

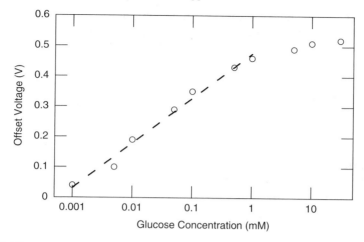

Figure 8.15 Dependence of the voltage offset on glucose concentration, where the voltage offset is the difference between the applied and effective gate voltage. The line is a guide to the eye with a slope of 147 mV per decade. Reprinted with permission from [44]. Copyright Royal Society of Chemistry.

equation that describes the dependence of chemical potential on the concentration of redox-active species:

$$E_{Nernst} = E^{0\prime} + \frac{k \cdot T}{n \cdot q} \cdot \ln\left(\frac{[Ox]}{[Red]}\right) \qquad (8.12)$$

where $[Ox]$ and $[Red]$ are the concentrations of oxidized and reduced species, $E^{0\prime}$ is the formal potential, k is the Boltzmann constant, T is the temperature, q the fundamental charge and n the number of electrons transferred during the reaction.

The physical meaning of the offset voltage can be understood by comparing sensors based on OECTs with conventional electrochemical sensors. In conventional electrochemistry, the effects of the Nernst equation are manifested by changes of the potential at a working electrode (where the reaction occurs) relative to a fixed reference electrode potential. In OECTs the gate voltage potential is fixed; consequently, the potential shift described by the Nernst equation is manifested by a shift of the electrolyte potential relative to the gate. In the non-Faradaic regime, the electrolyte potential is determined by the capacitances associated to the double-layer formation at the gate and channel and is equal to:

$$V_{sol} = \frac{V_g}{(1 + \gamma)} \qquad (8.13)$$

where γ is the capacitance ratio defined as $C_{channel}/C_{gate}$.

When glucose is added to an electrolyte solution containing GOx, the reaction shown in Figure 8.12a takes place and the potential drop across the Pt electrode/electrolyte interface decreases. This Faradaic contribution is described by the Nernst equation:

$$V^{sol} = \frac{V_g}{(1+\gamma)} + \frac{k \cdot T}{2 \cdot q} \cdot \ln[H_2O_2] + const \tag{8.14}$$

where the constant contains the details of proton and oxygen activity. This value of the electrolyte potential is described by the dashed line in Figure 8.16. From the equation above it is clear that the addition of glucose increases the electrolyte potential, which in turn decreases the drain current. It is convenient to introduce an effective gate voltage to describe this effect:

$$V_g^{eff} = V_g + (1+\gamma) \cdot \frac{k \cdot T}{2 \cdot q} \cdot \ln[H_2O_2] + const^* \tag{8.15}$$

where the new constant is that of Equation (8.13) multiplied by $(1+\gamma)$, V_g^{eff} is the equivalent gate voltage that needs to be applied in the absence of Faradaic effects to result in the same drain current. V_g^{eff} is illustrated in Figure 8.16 with a dotted line.

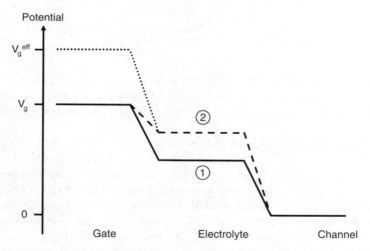

Figure 8.16   Diagram showing how the potential varies within an enzyme-based OECT. In the absence of reactions, the solution potential (1) is determined from the relative capacitances of the gate and channel. The solution potential in the Faradaic regime (2) is increased according to the Nernst equation. The effective gate voltage describes the required gate voltage to achieve the solution potential in the Faradaic regime in the absence of reactions. Reprinted with permission from [44]. Copyright Royal Society of Chemistry.

The above analysis also clarifies the physical meaning of the offset voltage involved in the transformation shown in Figure 8.16. V_{offset} is represented by the last two terms in Equation (8.15) and describes the Faradaic contribution to the effective gate voltage. It originates from the shift in the chemical potential described by the Nernst equation and is scaled by the capacitance ratio. The line in Figure 8.15 is a fit to V_{offset} with $\gamma = 4$. Given that the capacitance per unit area of conducting polymer electrodes is greater than that of metals, and that the area of the gate electrode was smaller than that of the channel, a value for γ that is larger than one is reasonable. It should be noted that the capacitance associated with metals and polymers is mechanistically distinct: while metals such as Pt are impermeable to ionic charge, ions can penetrate polymers, which gives rise to a unique origin for the capacitance in each.[45] The potential drop between the electrolyte and the channel in Figure 8.16 implies ion accumulation on the surface of the PEDOT:PSS. An effective capacitance can still be used for the case where ions completely penetrate the PEDOT:PSS.

Incorporation of the effective gate voltage in Equation (8.5) allows one to fit the sensor response (Figure 8.17). The normalized response (NR) of the drain current is plotted as a function of glucose concentration and gate voltage. Normalization was done relative to the zero concentration limit as:

$$NR = \frac{I_{sd}^{conc} - I_{sd}^{conc=0}}{I_{sd}^{conc=0}} \qquad (8.16)$$

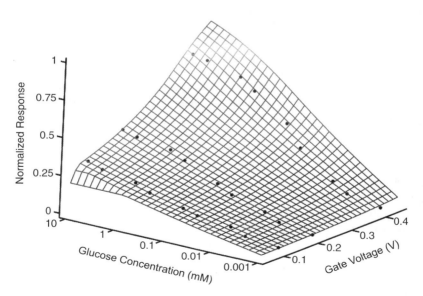

Figure 8.17 Plot of relative device response as a function of gate voltage and concentration. Points show experimental device response and surface shows the result of the model. Reprinted with permission from [44]. Copyright Royal Society of Chemistry.

where I_{sd} is considered at zero concentration and at the concentration of interest.

This normalization provides a maximum range of response from zero (no analyte) to one (upper concentration limit) and facilitates comparison between different devices. Figure 8.17 shows experimental data (filled circles) from PEDOT:PSS OECTs along with a fit to Equations (8.15)–(8.17), where $\gamma = 4$, $V_p = 0.8$ V and a correction for the resistivity of the source and drain electrodes is used.

8.4.2 Optimization of Device Performance

For operation as a remote sensor it is desirable to set up the experimental conditions so that the potential drop at the gate electrode/electrolyte interface is large. This will result in minimal initial gating of the transistor. The presence of the analyte will decrease the potential drop at the gate electrode/electrolyte interface and the gating of the transistor will increase. In terms of capacitance this translates into $C_g > C_{ch}$, which can easily be achieved by using a large gate electrode.

8.5 Conclusions and Perspectives

OECT-based sensors correspond to a new and exciting development in the field of organic electronics. These devices have been fabricated with different materials and have been used to detect a wide range of biological and chemical species. Modeling has provided an improved understanding that paves the way for rational device optimization. Ease of processing and low cost, together with tunability of electronic properties and integration with biological systems, make OECTs the ideal candidates for biosensing applications. However, the field is still in its infancy and many challenges and opportunities exist. To design and realize high-performance devices, the device physics of OECTs needs to be understood in more detail: for instance, by exploring how device performance depends on gate electrode material and size, on electrolyte chemistry and ionic strength and on device dimensions. Along this line it has been shown that the extent of gating in an OECT depends on the area of the gate electrode and on the amount of conducting polymer present in the channel.[46] Last but not least, one of the present limitations of OECT-based sensors is the limited number of available conducting polymers. Therefore new materials need to be developed in order to increase conductivity and improve long-term stability in aqueous environments.

Acknowledgements

This work was partially supported by the Nanoscale Science and Engineering Initiative of the National Science Foundation under NSF Award #EEC-0117770, 0646547 and the New York State Office of Science, Technology &

Academic Research under NYSTAR Contract # C020071. This work was performed in part at the Cornell NanoScale Facility, a member of the National Nanotechnology Infrastructure Network, which is supported by the National Science Foundation (Grant ECS-0335765). F. C. acknowledges the Marie Curie program of the European Union under project MC-OIF 040864-TOPOS for partial salary support.

Acronyms

BLM: bilayer lipid membrane
GOx: glucose oxidase
I_{sd}: drain current
NADH: nicotinamide adenine dinucleotide coenzyme
OECT: organic electrochemical transistor
OTFT: organic thin film transistor
PBS: phosphate buffered saline
PEDOT: poly(3,4-ethylenedioxythiophene)
PSS: poly(styrene sulfonate)
SCE: standard calomel (HgCl$_2$) electrode
V_d: drain voltage
V_g: gate voltage

References

1. G. G. Malliaras and R. H. Friend, *Phys. Today*, 2005, **58**, 53.
2. M. Berggren and A. Richter-Dahlfors, *Adv. Mater.*, 2007, **19**, 3201.
3. B. Crone, A. Dodabalapur, A. Gelperin, L. Torsi, H. E. Katz, A. J. Lovinger and Z. Bao, *Appl. Phys. Lett.*, 2001, **78**, 2229.
4. B. Choudhury, R. Shinar and J. Shinar, *J. Appl. Phys.*, 2004, **96**, 2949.
5. J. T. Mabeck and G. G. Malliaras, *Anal. Bioanal. Chem.*, 2006, **384**, 343.
6. H. S. White, G. P. Kittlesen and M. S. Wrighton, *J. Am. Chem. Soc.*, 1984, **106**, 5375.
7. S. H. Chao and M. S. Wrighton, *J. Am. Chem. Soc.*, 1987, **109**, 6627.
8. E. W. Paul, A. J. Ricco and M. S. Wrighton, *J. Phys. Chem.*, 1985, **89**, 1441.
9. J. W. Thackeray and M. S. Wrighton, *J. Phys. Chem.*, 1986, **90**, 6674.
10. P. N. Bartlett and P. R. Birkin, *Anal. Chem.*, 1993, **65**, 1118.
11. P. N. Bartlett and P. R. Birkin, *Anal. Chem.*, 1994, **66**, 1552.
12. M. Nishizawa, T. Matsue and I. Uchida, *Anal. Chem.*, 1992, **64**, 2642–2644.
13. T. Matsue, M. Nishizawa, T. Sawaguchi and I. Uchida, *Chem. Commun.*, 1991, **15**, 1029.
14. V. Saxena, V. Shirodkar and R. J. Prakash, *Solid State Electr.*, 2000, **4**, 234.

15. M. Kanungo, D. N. Srivastava, A. Kumar and A. Q. Contractor, *Chem. Commun.*, 2002, **7**, 680.
16. K. Krishnamoorthy, R. S. Gokhale, A. Q. Contractor and A. Kumar, *Chem. Commun.*, 2004, **7**, 820.
17. M. Nikolou and G. G Malliaras, *The Chemical Record*, in press.
18. S. Ashizawa, R. Horikawa and H. Okuzaki, *Synth. Met.*, 2005, **153**, 5.
19. J. Ouyang, Q. F. Xu, C. W. Chu, Y. Yang, G. Li and J. Shinar, *Polymer*, 2004, **42**, 8443.
20. J. Y. Kim, J. H. Jung, D. E. Lee and J. Joo, *Synth. Met.*, 2002, **126**, 311.
21. D. Nilsson, N. Robinson, M. Berggren and R. Forchheimer, *Adv. Mater.*, 2005, **17**, 353.
22. D. Nilsson, M. Chen, T. Kugler, T. Ramonen, M. Armgarh and M. Berggren, *Adv. Mater.*, 2002, **14**, 51.
23. J. Isaksson, C. Tengstedt, M. Fahlman, N. Robinson and M. Berggren, *Adv. Mater.*, 2004, **16**, 316.
24. D. Nilsson, T. Kugler, P. O. Svensson and M. Berggren, *Sens. Act. B*, 2002, **86**, 193.
25. J. Isaksson, P. Kjall, D. Nilsson, N. D. Robinson, M. Berggren and A. Richter-Dahlfors, *Nature Mater.*, 2007, **6**, 673.
26. K. Krishnamoorthy, R. S. Gokhale, A. Q. Contractor and A. Kumar, *Chem. Commun.*, 2004, 820.
27. A. Q. Contractor, A. Kumar, R. Narayanan, S. Sukeerthi, R. Lal and R. S. Srinivas, *Electrochimica Acta*, 1994, **39**, 1321.
28. H. Sangodkar, S. Sukeerthi, R. S. Srinivas, R. Lal and A. Q. Contractor, *Anal. Chem.*, 1996, **68**, 779.
29. S. Sukeerthi and A. Q. Contractor, *Anal. Chem.*, 1999, **71**, 2231.
30. M. Kanungo, A. Kumar and A. Q. Contractor, *Anal. Chem.*, 2003, **75**, 5673.
31. R. B. Dabke, G. D Singh, A. Dhanabalan, R. Lal and A. Q. Contractor, *Anal. Chem.*, 1997, **69**, 724.
32. D. Raffa, K. T. Leung and F. Battaglini, *Anal. Chem.*, 2003, **75**, 4983.
33. P. N. Bartlett, P. R. Birkin, J. H. Wang, F. Palmisano and G. D. Benedetto, *Anal. Chem.*, 1998, **70**, 3685.
34. Z. T. Zhu, J. T. Mabeck, C. Zhu, N. C. Cady, C. A. Batt and G. G. Malliaras, *Chem. Commun.*, 2004, 1556.
35. D. A. Bernards, G. G. Malliaras, G. E. S. Toombes and S. M. Gruner, *Appl. Phys. Lett.*, 2006, **89**, 053505.
36. J. T. Mabeck, J. A. DeFranco, D. A. Bernards, G. G. Malliaras, S. Hocde and C. J. Chase, *Appl. Phys. Lett.*, 2005, **87**, 1.
37. D. A. Bernards and G. G. Malliaras, *Adv. Funct. Mater.*, 2007, **17**, 3538.
38. A. J. Bard and L. R. Faulkner, *Electrochemical Methods*, Wiley, New York, 1980.
39. J. D. Stenger-Smith, C. K. Webber, N. Anderson, A. P. Chafin, K. Zong and J. R. Reynolds, *J. Electrochem. Soc.*, 2002, **149**, A973.
40. W. Shockley, *Proc. IRE*, 1952, **40**, 1365.

41. M. M. Alam, J. Wam, Y. Guo, S. P. Lee and H. R. Tseng, *J. Phys. Chem. B*, 2005, **109**, 12777.
42. J. T. Wallmark, *RCA Rev.*, 1963, **24**, 641.
43. H. T. Tien and A. Ottowa-Leitmannova, ed., *Planar Lipid Membranes and their Applications*, Elsevier, Amsterdam, 2003.
44. D. A. Bernards, G. G. Malliaras, D. J. Macaya, M. Nikolu, J. A. DeFranco, S. Takamatsu and G. G. Malliaras, *J .Mater. Chem.*, 2008, **18**, 116.
45. J. Wang and A. Bard, *J. Am. Chem. Soc.*, 2001, **123**, 498.
46. F. Lin and M. Lonergan, *Appl. Phys. Lett.*, 2006, **88**, 133507.

Subject Index